W0037671

Introduction to Supersymmetry in Particle and Nuclear Physics

Introduction to Supersymmetry in Particle and Nuclear Physics

Edited by

O. Castaños, A. Frank, and L. Urrutia

Centro de Estudios Nucleares
Universidad Nacional Autonoma de Mexico
Mexico City, Mexico

Springer Science+Business Media, LLC

Library of Congress Cataloging in Publication Data

International School of Supersymmetry (1981: Mexico City, Mexico)
Introduction to supersymmetry in particle and nuclear physics.

"Proceedings of an International School of Supersymmetry, held December 14–18, 1981, in Mexico City, Mexico"—T.p. verso.
Includes bibliographical references and index.
1. Supersymmetry—Congresses. I. Castaños, O. II. Frank, A. III. Urrutia, L. IV. Title.
QC174.17.S9I58 1981 539.7′2 83-26943
ISBN 978-1-4757-0919-3 ISBN 978-1-4757-0917-9 (eBook)
DOI 10.1007/978-1-4757-0917-9

Proceedings of an International School of Supersymmetry,
held December 14–18, 1981,
in Mexico City, Mexico

PREFACE

In the fall of 1981 the Centro de Estudios Nucleares of the Universidad Nacional Autónoma de México organized an International School of Supersymmetry which took place in Mexico City from Dec. 14 to Dec. 18.

The purpose of this School was to provide both students and researchers with an introduction to Supersymmetry as well as an overview of current research topics.

A general introduction to the subject was given by Dr. Freedman while Dr. Grisaru discussed the superspace formulation of Supersymmetry and Supergravity. Applications of these ideas to Particle Physics were discussed by Dr. Ferrara and Dr. Witten. Finally, Dr. Bars presented the basic framework for the discussion of Supersymmetries in Nuclear Physics.

We would like to take this opportunity to thank our lecturers for their enthusiastic participation in the School.

The collaboration of Dr. Marcos Rosenbaum, Director of the Centro de Estudios Nucleares, and of all our colleagues in the Physics Department is also gratefully acknowledged.

We also thank Mrs. Ma. Esther Colmenares for her careful typing of the manuscript and also for her unlimited patience in making all the necessary corrections. The help of Mr. Jose Rangel with symbols and figures is also deeply appreciated.

O. Castaños
A. Frank
L. Urrutia

CONTENTS

INTRODUCTION TO SUPERSYMMETRY

Daniel Z. Freedman

Department of Mathematics
Massachusetts Institute of Technology
Cambridge, Massachusetts 02139

One theme which underlies the development of particle physics is the application of ever broader symmetries to particle interactions. A new symmetry always relates and unifies previously distinct phenomena. The initial appeal of a symmetry may be based only on reasons of mathematical elegance. It may take some years to find applications in nature.

The traditional symmetries of quantum field theories are the space-time symmetry of the Poincaré group and internal symmetries such as the SU(3) group of strong interaction flavor and color. Supersymmetry is a larger symmetry which must contain the Poincaré group and can contain internal symmetry also, thus unifying them.

This unification is accomplished by bringing the fundamental quantum mechanical notions of spin and statistics into the formulation of the symmetry. Above all, supersymmetry relates particles of different spin and relates bosons and fermions. There are theorems which indicate that supersymmetry is the broadest symmetry possible in relativistic quantum field theory.

These lectures are intended to be an introduction to supersymmetry which conveys the general scope of the subject and some of its specific technical formalism. In discussing supersymmetric field theories, we will use the "component approach", which is designed to bring out the physical content with a minimum of formalism. Other approaches such as superspace (discussed by Dr. Grisaru) are better suited to bring out the geometric structure of supersymmetry and are necessary for a deeper understanding of the physics. However, at the moment, the component approach is the only way known in which all supersymmetric theories can be formulated. The superspace formalism or the (roughly equivalent) auxiliary field formulation is not yet known for some of the more interesting theories.

Symmetries in Relativistic Quantum Field Theory

We will now survey the various continuous symmetries used in particle physics, discussing: i) the nature of the conserved charges and the algebraic system they determine; ii) what is unified, by which we mean what kinds of particle states are brought together in irreducible representations of the algebra; iii) the conservation laws and other features which arise when the symmetry operates at the global level (parameters independent of space-time coordinates x^μ); and iv) the special gauge fields which are required when the symmetry operates at the local level (with parameters depending on x^μ); and v) some applications.

The Poincaré group is the group of space-time symmetries and contains vector charges P_μ which generate translations and antisymmetric tensor charges $M_{\mu\nu}$ which generate rotations and Lorentz transformations. These charges determine a Lie algebra with commutation relations

$$
\begin{aligned}
\left[M_{\mu\nu}, M_{\lambda\rho}\right] &= g_{\nu\lambda} M_{\mu\rho} - g_{\mu\lambda} M_{\nu\rho} - g_{\nu\rho} M_{\mu\lambda} + g_{\mu\rho} M_{\nu\lambda} , \\
\left[M_{\mu\nu}, P_\rho\right] &= g_{\nu\rho} P_\mu - g_{\mu\rho} P_\nu , \qquad (1) \\
\left[P_\mu, P_\nu\right] &= 0 ,
\end{aligned}
$$

where $g_{\mu\nu}$ is the Minkowski metric.

An important implication of Poincaré invariance is that particles are classified by mass and spin, and that the various momentum and helicity states $|\vec{p},\lambda\rangle$ of a single particle, such as the proton, are unified in an irreducible representation. Global Poincaré invariance predicts that energy, momentum, and angular momentum are conserved in any reaction. It also implies restriction on scattering amplitudes; for example, two-body scattering amplitudes $A(s,t)$ depend only on two variables, the energy and momentum transfer invariants. In any theory with local Poincaré invariance, a massless spin 2-gauge field appears, and one finds a theory of gravitation, normally a variant of general relativity.

Next we have internal symmetries with scalar (and pseudoscalar) charges T^a and commutation relations

$$\left[T^a, T^b\right] = f^{abc}\, T^c , \qquad (2)$$

of some compact Lie algebra. Particles of the same mass and spin (and fixed momentum and helicity) but different eigenvalues of the charges are unified in irreducible representations with basis $|\vec{p},\lambda,i\rangle$ where i denotes the internal quantum numbers (or eigenvalues of commuting T^a). A global internal symmetry predicts conservation laws for additive quantum numbers and relations among amplitudes for different reactions obtained from Clebsch-Gordan analysis. Gauge internal symmetry requires spin 1 gauge fields, and we know that this leads to successful models of unified weak and electromagnetic interactions (the Weinberg-Salam Model), for strong interactions (QCD), and further prospects for unification of all three.

Before going on to supersymmetry, let us digress and note that an important consequence of the Coleman-Mandula theorem is that any Lie group symmetry in interacting field theory must factor into a direct product of the Poincaré group and an internal symmetry group. Any further unification of spin and internal symmetry requires a different mathematical framework than Lie algebras. Fortunately, there is the formalism of graded Lie algebras, which are algebraic systems with two classes (or grades) of elements, the even elements B^a and odd elements F_α and structure relations of the form

$$\begin{aligned} \left[B^a,B^b\right] &= f^{abc}\, B^c \;, \\ \left[B^a,F_\alpha\right] &= -t^a{}_{\alpha\beta}\, F_\beta \;, \\ \left\{F_\alpha,F_\beta\right\} &= s_{a\alpha\beta}\, B^a \;. \end{aligned} \tag{3}$$

Thus the even elements determine a Lie algebra, the odd elements transform in a representation of that algebra with representation matrices $t^a{}_{\alpha\beta}$ and the anticommutator of two odd elements is a linear combination of even elements. One can easily derive graded Jacobi identities which place restrictions on t and s, given f. Using these relations, graded Lie algebras have been completely classified with results analogous to the Cartan classification of Lie algebras. (For more detail, see the lectures of Dr. Bars.) We will not discuss this classification because relativistic supersymmetries are restricted to a small class whose structure will be discussed in physical terms. One should note that graded Lie algebras are not mysterious systems, they are very concrete and can always be represented by block diagonal matrices

$$B = \begin{bmatrix} b & o \\ o & \tilde{b} \end{bmatrix} \;, \qquad F = \begin{bmatrix} o & f \\ \tilde{f} & o \end{bmatrix} \;, \tag{4}$$

with dimension nxn for b, mxm for $\tilde{b}$, nxm for f, and mxn for $\tilde{f}$.

The supersymmetry algebra is a graded extension of the Poincaré group in which one or more spinor charges appear as odd elements. First let us take the simplest case of one such charge Q_α with four components $\alpha = 1\ldots4$. The structure relations are those given in

(1) for M and P together with

$$[M^{\mu\nu}, Q_\alpha] = -i(\sigma^{\mu\nu})_{\alpha\beta}\, Q_\beta \,, \tag{5}$$

$$[P_\mu, Q_\alpha] = 0 \,, \tag{6}$$

$$\{Q_\alpha, \overline{Q}_\beta\} = (\gamma^\mu)_{\alpha\beta}\, P_\mu \,, \tag{7}$$

which specify the graded segment of the algebra. The structure constants $t^a{}_{\alpha\beta}$ and $s_{a\alpha\beta}$ of (3) are essentially Dirac matrices γ^μ with $\{\gamma^\mu, \gamma^\nu\} = 2g^{\mu\nu}$ and $\sigma^{\mu\nu} = 1/4\,[\gamma^\mu, \gamma^\nu]$. Eq. (5) simply states that Q_α transforms as a spinor under Lorentz transformations, while (6) implies that the spinor charges are conserved and translation invariant. Eq. (7) is the key structure relation of supersymmetry. One might expect an anti-commutator because of the connection between spin and statistics. The heart of the matter is that repeated supersymmetry transformations give a translation in space-time, so that this familiar operation can be factored into more elementary ones.

To see what particles are unified in supersymmetry, consider a helicity state $|\vec{p}, \lambda\rangle$ and apply the charge Q_α. The result is

$$Q_\alpha\, |\vec{p}, \lambda\rangle = a\,|\vec{p}, \lambda+1/2\rangle + b\,|\vec{p}, \lambda-1/2\rangle \;. \tag{8}$$

Since $[P, Q] = 0$, the right side must be a superposition of particles of the same momentum and energy: therefore the same mass. Addition of angular momenta implies that these particles have helicities $\lambda \pm 1/2$. Thus supersymmetry transformations connect particle states which differ by 1/2 unit of spin. Hence they relate bosons and fermions.

A more precise statement of what particles are unified by supersymmetry is that bosons and fermions of equal mass are joined in irreducible representations of the algebra. For massless particles these representations consist of adjacent spin boson-fermion doublets s, s-1/2 which exist for all quantum mechanical spin values s = 1/2,1,... . For $m \neq 0$, a representation requires four particles of spins s, s-1/2, s-1/2, s-1 which again exist for all s = 1/2, 1,... .

In global supersymmetry, one has conservation laws and relations between amplitudes analogous to those of Poincaré invariance and internal symmetry. The conservation laws are purely formal since eigenstates of Q_α would be superspositions of bosons and fermions which are prohibited by superselection rules. One does have relations between scattering amplitudes for reactions involving different numbers of bosons and fermions, for example, the equality $\langle 1, 1/2|S|1, 1/2\rangle = \langle 1/2, 1/2|S|1/2, 1/2\rangle$ for helicity amplitudes in the spin 1, spin 1/2 system.

In local supersymmetry, the (2,3/2) system is singled out as the gauge multiplet. This leads to the theory of supergravity where the spin 3/2 gravitino shares the stage with the spin 2 graviton.

The graded Lie algebra just discussed is called the N = 1 Poincaré supersymmetry algebra. One can accommodate a global or gauge internal symmetry group G as a direct product, but there is no unification with supersymmetry. Later we will discuss the important extended algebras where there are N such spinor charges and unification with internal symmetry. At the moment, there is an active line of research to develop N = 1 global supersymmetric theories which apply to grand unification of the weak, electromagnetic and strong interactions, as discussed by Dr. Witten. The major application of extended supersymmetry appears to be in supergravity, where one hopes for a truly unified field theory of all known forces.

Supersymmetry in Quantum Field Theory

It is important now to show that the algebraic structure discussed above can be realized in interacting quantum field theory. It should be stressed that supersymmetry always acts as a conventional Lagrangian symmetry with the following features.

a) We deal with systems of boson fields $\phi(x)$ and fermion fields $\psi(x)$ (Lorentz and spin indices are suppressed) with local Lagrangian density and action integral

$$I[\phi,\psi] = \int d^4x\, L(\phi,\psi,\partial_\mu\phi,\partial_\mu\psi) \; . \tag{9}$$

b) There is an anticommuting spinor parameter ε_α, and supersymmetry variations of the fields which have the schematic form

$$\begin{aligned} \delta\phi &= \bar{\varepsilon}\Gamma\psi \; , \\ \delta\psi &= \partial\phi\,\tilde{\Gamma}\varepsilon + \ldots \; , \end{aligned} \tag{10}$$

where Γ and $\tilde{\Gamma}$ are Dirac matrices chosen to balance Lorentz indices, and +... indicates possible additional terms. The transformations always mix bose and fermi fields.

c) The action is invariant under these transformations, but the Lagrangian density always changes by an explicit divergence

$$\delta L = \partial_\mu K^\mu(\phi,\psi,\varepsilon) \; , \tag{11}$$

which is non-vanishing because supersymmetry is like a space-time symmetry and a partial integration is necessary to show invariance. The Noether current

$$J^{\mu} = \frac{\delta L}{\delta \partial_{\mu}\phi}\,\delta\phi + \frac{\delta L}{\delta \partial_{\mu}\psi}\,\delta\psi - K^{\mu} \quad , \tag{12}$$

is conserved and typically bilinear in bose and fermi fields. The charges

$$\bar{\varepsilon}_{\alpha}Q_{\alpha} = \int d^3x\, J^0 \quad , \tag{13}$$

then generate the supersymmetry algebra, when acting on particle states of the theory. The question of the algebra obtained by repeated supersymmetry transformations on the fields is more delicate and not without importance, but we will not discuss it at this point.

d) The dynamics of the particular supersymmetric field theory determines if the symmetry is realized either as

i) an exact symmetry with physical particle states in representations of the algebra with boson-fermion mass equality, or as

ii) a spontaneously broken symmetry with $m_{Bose} \neq m_{Fermi}$ with a massless Goldstone fermion in a global theory or a massive spin 3/2 Higgs particle in supergravity.

We will discuss the field theories using the language of exact supersymmetry. However, one must bear in mind that supersymmetry is broken in nature if it is true at all. The known dynamical mechanisms for breakdown are discussed in the lectures of Dr. Ferrara.

e) A simple dimensional argument can be made which suggests the relation between supersymmetry and translations. Since the canonical dimensions of ϕ and ψ are 1 and 3/2, respectively, we see from $\delta\phi$ in (10) that ε has dimension -1/2. This means that in $\delta\psi = ?\ \phi\tilde{\Gamma}\varepsilon$ we must supply one dimensional unit, and the choice is either differentiation or multiplication by a dimensional parameter. In fundamental massless theories, only differentiation is available and always appears as we have indicated in (10). Because of the derivative, $\{Q,\bar{Q}\}$ contains the translation P_{μ}.

Dirac Matrices and Majorana Spinors

Now we need some simple but technical material on Dirac matrices and Majorana spinors. The motivation is that to build up supersymmetry one should work with spinors which are most fundamental in the sense of having the fewest independent components. This means either Weyl spinors with two complex components or Majorana spinors with four independent real components. The two formalisms are equivalent, and we choose the latter.

In the Dirac algebra generated by γ^μ matrices (which satisfy $\{\gamma^\mu,\gamma^\nu\} = 2g^{\mu\nu}$), there is a charge conjugation matrix C which satisfies $C\gamma^\mu C^{-1} = -(\gamma^\mu)^T$ where T indicates transpose. Independent of representation, one can show that the 6 matrices C, $C\gamma_5$ and $C\gamma_5\gamma^\mu$ are antisymmetric, and the 10 matrices $C\gamma^\mu$ and $C\sigma^{\mu\nu}$ are symmetric. Further, the spinors ψ and $C^{-1}\bar{\psi}^T$ transform in the same way under Lorentz transformations. In most representations, we can choose $C = -C^{-1} = C^*$ (or $C = C^{-1} = -C^*$ after multiplication by the constant i). Given any four component spinor ψ, it is then convenient to define its charge conjugate $\psi^C = C\bar{\psi}^T = C\gamma^{0T}\psi^*$, and one has $\psi^{CC} = \psi$. One can then impose the Lorentz invariant self-conjugacy condition

$$\psi = \psi^C = C\bar{\psi}^T , \tag{14}$$

which defines a Majorana spinor. This condition is analogous to the reality condition $\phi = \phi^*$ for bose fields, and clearly halves the number of independent components of $\psi(x)$. Indeed, in a Majorana representation, defined by $C\gamma^{0T} = 1$, we have $\psi = \psi^*$ so that a Majorana spinor is purely real. In the signature (+,−,−,−), the γ-matrices of a Majorana representation are all imaginary. One such representation is given by

$$\gamma^0 = \begin{bmatrix} \sigma_2 & 0 \\ 0 & -\sigma_2 \end{bmatrix} , \qquad \gamma^1 = \begin{bmatrix} i\sigma_1 & 0 \\ 0 & -i\sigma_1 \end{bmatrix} ,$$

$$\gamma^2 = \begin{bmatrix} 0 & i \\ i & 0 \end{bmatrix} , \qquad \gamma^3 = \begin{bmatrix} i\sigma_3 & 0 \\ 0 & i\sigma_3 \end{bmatrix} ,$$

$$C = \gamma^{0T} . \tag{15}$$

In a representation with γ_5 diagonal, such as

$$\gamma^0 = \begin{bmatrix} 0 & 1 \\ 1 & 0 \end{bmatrix} , \qquad \gamma^i = \begin{bmatrix} 0 & \sigma_i \\ -\sigma_i & 0 \end{bmatrix} ,$$

$$\gamma_5 = i\gamma^0\gamma^1\gamma^2\gamma^3 = \begin{bmatrix} -1 & 0 \\ 0 & 1 \end{bmatrix} , \qquad C = i\gamma^2\gamma^0 , \tag{16}$$

a Majorana spinor takes the form

$$\psi_\alpha = \begin{bmatrix} \psi_1 \\ \psi_2 \\ \psi_2{}^* \\ -\psi_1{}^* \end{bmatrix} , \tag{17}$$

and upper (and lower) components transform irreducibly under Lorentz transformations as left-handed (right-handed) Weyl spinors.

Until further notice, all spinor quantities, including the charges Q_α of supersymmetry algebras, will be assumed to be Majorana spinors. Given any two Majorana spinors, ψ and χ, one has $\bar{\psi}\Gamma\chi = \psi^T C\Gamma\chi$ for any Dirac matrix Γ. We now consider the two choices of commuting or anticommuting numbers, for the components of these spinors, viz.,

$$\chi_\alpha \psi_\beta \mp \psi_\beta \chi_\alpha = 0 . \tag{18}$$

Using the symmetry properties stated earlier, one then has

$$\begin{aligned} \bar{\psi}\chi &= \mp \bar{\chi}\psi , \\ \bar{\psi}\gamma^\mu\chi &= \pm \bar{\chi}\gamma^\mu\psi , \\ \bar{\psi}\sigma^{\mu\nu}\chi &= \pm \bar{\chi}\sigma^{\mu\nu}\psi , \\ \bar{\psi}\gamma_5\gamma^\mu\chi &= \mp \bar{\chi}\gamma_5\gamma^\mu\psi , \\ \bar{\psi}\gamma_5\chi &= \mp \bar{\chi}\gamma_5\psi . \end{aligned} \tag{19}$$

Let us study free field theory for Majorana spinors. The unique Lorentz and parity invariant bilinear action is

$$I = 1/2 \int d^4x \, \bar{\psi}(x)(i\not{\partial} - m)\, \psi(x) . \tag{20}$$

Of course, this is just the standard Dirac action, which is multiplied by 1/2 because a Majorana field has 1/2 the number of independent components. Now we notice that I vanishes identically if the components of ψ_α commute, and we are therefore forced to take anticommuting numbers, the lower sign in (18) and (19), even at the prequantum level. (This question does not arise for complex Dirac fields which can always be written as a sum of two Majorana fields, $\psi = 1/\sqrt{2}\,(\psi^1 + i\psi^2)$, because for both assignments of statistics either the diagonal $\psi^1\psi^1 + \psi^2\psi^2$ or off-diagonal $\psi^1\psi^2 - \psi^2\psi^1$ terms contribute).

To find the equations of motion, one takes the first variation of (20), according to the procedure

$$\delta I = \tfrac{1}{2} \int d^4x \left[\delta\bar{\psi}\,(i\not\partial - m)\psi + \bar{\psi}(i\not\partial - m)\,\delta\psi\right] ,$$

$$= \int d^4x\ \delta\bar{\psi}\,(i\not\partial - m)\psi , \tag{21}$$

where we have used partial integration and the Majorana property in the second step. Thus, in effect it is necessarily only to vary $\bar{\psi}$ and multiply by 2, a simplification which holds for all global supersymmetric theories (but must be used cautiously in supergravity because covariant derivatives cannot always be partially integrated). The equation of motion is just the Dirac equation $(i\not\partial - m)\psi = 0$, which comes as no surprise, and the plane wave expansion of the general solution is

$$\psi(x) = \int \frac{d^3p}{(2\pi)^{3/2}(2\omega)^{1/2}} \sum_{\lambda=\pm 1/2} \left[b(\vec{p},\lambda)u(\vec{p},\lambda)e^{-ip\cdot x}\right.$$

$$\left. + b^*(\vec{p},\lambda)v(\vec{p},\lambda)e^{ip\cdot x}\right] . \tag{22}$$

One sees that the usual u and v spinors appear, but that the Fourier coefficients of positive and negative frequency waves are conjugate as required by the Majorana condition. After quantization, one has

$$\{b(p,\lambda)\ ,\ b^*(p',\lambda')\} = \hbar\ \delta^3(p - p')\delta_{\lambda\lambda'} , \tag{23}$$

and the b and b^* become annihilators and creators of particle states which can be described as follows. For each momentum $\vec{p}$, one has two particles of helicity $\lambda = \pm 1/2$. Particles are identical with antiparticles, as for real boson fields.

Supersymmetric Yang-Mills Theory

We now consider the supersymmetric Yang-Mills theory which is an elegant global supersymmetric theory (and should convince you that there is some depth in the subject). This is the field theory of the (1,1/2) massless representation of supersymmetry, in direct product with an arbitrary internal symmetry group G. The fields of the model are a set of vector potentials $A_\mu{}^a(x)$ and their fermionic partners, a set of Majorana fermion fields $\chi^a(x)$. Both are assigned to the adjoint representation of G.

The supersymmetric action of this system is simply the minimal-coupled action in the sense of Yang-Mills invariance, namely

$$I[A_\mu{}^a,\chi] = \int d^4x\ \{-\tfrac{1}{4}(F_{\mu\nu}{}^a)^2 + 1/2\ i\ \bar{\chi}^a\gamma^\mu(D_\mu\chi)^a\} ,$$

$$F_{\mu\nu}{}^a = \partial_\mu A_\nu{}^a - \partial_\nu A_\mu{}^a + g\ f^{abc}\ A_\mu{}^b\ A_\nu{}^c$$

$$(D_\mu\chi)^a = \partial_\mu\chi^a + g\, f^{abc}\, A_\mu{}^b\, \chi^c \ . \tag{24}$$

One could have written down this theory well before the development of supersymmetry without suspecting that it has an additional fermionic invariance, which is now known to be given by the transformation

$$\begin{aligned} \delta A_\mu{}^a &= i\, \bar{\varepsilon}\gamma_\mu\chi^a \ , \\ \delta\chi^a &= \sigma^{\mu\nu}\, F_{\mu\nu}{}^a\, \varepsilon \ , \\ \Rightarrow \delta\bar{\chi}^a &= -\bar{\varepsilon}\, \sigma^{\mu\nu}\, F_{\mu\nu}{}^a \ . \end{aligned} \tag{25}$$

This has the same general character discussed earlier; bose fields rotate into fermi fields; fermi fields rotate into derivatives of the bose fields.

We will give a detailed proof of invariance as an example of the kind of manipulation which occurs frequently in the component approach to supersymmetry and supergravity. First compute the variation of the action, taking advantage of the simplification made earlier concerning variation of χ only, valid here because Yang-Mills derivatives can be partially integrated. Thus

$$\begin{aligned} \delta I &= \int d^4x\, \left\{\delta A_\mu{}^a \frac{\delta I}{\delta A_\mu{}^a} + \delta\chi^a \frac{\delta I}{\delta\chi^a}\right\} \ , \\ &= \int d^4x \{ i\bar{\varepsilon}\gamma_\mu\chi^a \left[(D^\nu F_{\mu\nu})^a + 1/2\ ig\, f^{abc}\, \bar{\chi}^b\gamma^\mu\chi^c \right] \\ &\qquad - i\bar{\varepsilon}\, \sigma^{\alpha\beta}\, F_{\alpha\beta}{}^a\, \gamma^\mu (D_\mu\chi)^a \} \ . \end{aligned} \tag{26}$$

The Yang-Mills and Dirac equations of motion appear as factors, as they must, but one cannot assume that the equations are satisfied, since a Lagrangian symmetry requires that the action be stationary for all field configurations, not just those which satisfy the equations of motion. Now integrate the last term by parts and change the order of terms to write

$$\begin{aligned} \delta I = \int d^4x \{ & i\bar{\varepsilon}\gamma^\mu\chi^a (D^\nu F_{\nu\mu})^a + i\bar{\varepsilon}\, \sigma^{\alpha\beta}\, \gamma^\mu (D_\mu F_{\alpha\beta})^a\, \chi^a \\ & - 1/2\ g\, f^{abc} (\bar{\varepsilon}\gamma^\mu\chi^a)(\bar{\chi}^b\gamma_\mu\chi^c) \} \ . \end{aligned} \tag{27}$$

Now use the Dirac matrix identity

$$\begin{aligned} \sigma^{\alpha\beta}\, \gamma^\mu &= 1/2\left[\sigma^{\alpha\beta},\gamma^\mu\right] + 1/2\{\sigma^{\alpha\beta},\gamma^\mu\} \ , \\ &= 1/2(\gamma^\alpha\, g^{\beta\mu} - \gamma^\beta\, g^{\alpha\mu}) + 1/2\ i\varepsilon^{\alpha\beta\mu\nu}\, \gamma_5\gamma_\nu \ , \end{aligned} \tag{28}$$

to rewrite the second term in (27) as

$$\delta I_2 = i\int d^4x\ \bar{\varepsilon}[\gamma^{\alpha}(D^{\beta}F_{\beta\alpha})^a - i\gamma_5\gamma_{\nu}\ \varepsilon^{\nu\mu\alpha\beta}(D_{\mu}F_{\alpha\beta})^a]\chi^a \ . \quad (29)$$

The last term vanishes as a consequence of the gauge field Bianchi identity and the first term in (29) cancels the first term in (27).

At this point we are left with

$$\delta I = -1/2\ g\int d^4x\ f^{abc}(\bar{\varepsilon}\gamma^{\mu}\chi^a)(\bar{\chi}^b\gamma_{\mu}\chi^c) \quad (30)$$

and we will now show that the integrand vanishes due to properties of the Dirac matrices and anticommutation rules for the spinors. Since the 16 Dirac matrices Γ^A are a complete set of 4x4 matrices, (where A is an index which ranges through 16 values) we can write

$$(\Gamma^A)_{\alpha\beta}(\Gamma^B)_{\gamma\delta} = M^{AB;CD}(\Gamma^C)_{\alpha\delta}(\Gamma^D)_{\gamma\beta}\ , \quad (31)$$

where $M^{AB;CD}$ is a 256x256 numerical matrix which is the Fierz matrix in its most general form. Fortunately, in practice one is interested only in Lorentz-invariant quantities with simpler rearrangement properties, such as

$$(\gamma^{\mu})_{\alpha\beta}(\gamma_{\mu})_{\gamma\delta} = a\delta_{\alpha\delta}\delta_{\gamma\beta} + b(\gamma^{\mu})_{\alpha\delta}(\gamma_{\mu})_{\gamma\beta} + c(\sigma^{\mu\nu})_{\alpha\delta}(\sigma_{\mu\nu})_{\gamma\beta}$$
$$+ d(\gamma_5\gamma^{\mu})_{\alpha\delta}(\gamma_5\gamma_{\mu})_{\gamma\beta} + e(\gamma_5)_{\alpha\delta}(\gamma_5)_{\gamma\beta}\ . \quad (32)$$

By "tracing" with matrices $\delta_{\beta\alpha}$, $(\gamma^{\rho})_{\beta\alpha}$, etc., one easily finds the coefficients a = 1, b = −1/2, c = 0, d = −1/2, e = −1. Thus for any three Majorana spinors ψ^1, ψ^2, ψ^3 one has

$$\gamma^{\mu}\psi^1(\bar{\psi}^2\gamma_{\mu}\psi^3) = \pm\ [\psi^3(\bar{\psi}^2\psi^1) - 1/2\gamma^{\mu}\psi^3(\bar{\psi}^2\gamma_{\mu}\psi^1)$$
$$- 1/2\gamma_5\gamma^{\mu}\psi^3(\bar{\psi}^2\gamma_5\gamma^{\mu}\psi^1) - \gamma_5\psi^3(\bar{\psi}^2\gamma_5\psi^1)]\ , \quad (33)$$

where ± refers to commuting and anti-commuting statistics, respectively. Since the anti-commuting choice has already been made for supersymmetry, we apply this to the integrand of (33) as

$$f^{abc}(\bar{\varepsilon}\gamma^{\mu}\chi^a)(\bar{\chi}^b\gamma_{\mu}\chi^c) = -f^{abc}[(\bar{\varepsilon}\chi^c)(\bar{\chi}^b\chi^a) - 1/2(\bar{\varepsilon}\gamma^{\mu}\chi^c)(\bar{\chi}^b\gamma_{\mu}\chi^a)$$
$$-1/2(\bar{\varepsilon}\gamma_5\gamma^{\mu}\chi^c)(\bar{\chi}^b\gamma_5\gamma_{\mu}\chi^a) - (\bar{\varepsilon}\gamma_5\chi^c)(\bar{\chi}^b\gamma_5\chi^a)]\ . \quad (34)$$

Due to anti-symmetry $f^{abc} = -f^{bac}$ and the Majorana exchange properties (19), all terms except $\gamma^{\mu}\gamma_{\mu}$ vanish on the right side of (34) and we can write (after an internal symmetry index change)

$$f^{abc}(\bar{\varepsilon}\gamma^{\mu}\chi^a)(\bar{\chi}^b\gamma_{\mu}\gamma^c) = -1/2\ f^{abc}(\bar{\varepsilon}\gamma^{\mu}\chi^a)(\bar{\chi}^b\gamma_{\mu}\chi^c)\ . \quad (35)$$

Thus, the integrand in (30) vanishes, basically because of over-antisymmetrization of spinors, and we have finally proven that the field theory is supersymmetric.

The elements of the proof of invariance are:

a) Structure of the Dirac algebra, including Fierz rearrangement;

b) Anti-commuting statistics of fermion fields;

c) Basic identities of the other invariance of the problem, i.e., the Bianchi identity of Yang-Mills gauge invariance.

In supergravity, fundamental identities of the curvature tensor are required. These elements recur universally in the component approach to supersymmetry and reflect the intricate connection between spin, statistics, space-time structure, and internal symmetry, which supersymmetry embodies.

Scalar Multiplet and Auxiliary Fields

We now discuss field theories of the (1/2, 0) representation of the supersymmetry algebra. The fields of these theories form what is known as the scalar or chiral multiplet, and the theories are generalizations of the Wess-Zumino model. We consider n-multiplets of fields

$$
\begin{aligned}
&Z^a = 1/\sqrt{2}\ (A^a + B^a) \ , \\
&\psi^a \ , \qquad\qquad\qquad\qquad a = 1,\ldots,n \ , \qquad\qquad (36) \\
&F^a = 1/\sqrt{2}\ (\tilde{F}^a + \tilde{G}^a) \ ,
\end{aligned}
$$

where Z^a is a complex scalar whose real and imaginary parts are scalar and pseudoscalar, ψ^a is a Majorana spinor, and F^a is a complex scalar auxiliary field. The kinetic Lagrangian of this multiplet is

$$L_{kin} = \partial_\mu \bar{Z}^a \partial_\mu Z^a + 1/2\ i\bar{\psi}^a \not{\partial} \psi^a + \bar{F}^a F^a \ , \qquad (37)$$

where $\bar{Z}^a, \bar{F}^a$ indicate complex conjugate, and there is a sum on repeated indices. It is clear that Z and ψ are propagating fields while F is non-propagating. Part of the purpose of this discussion is to explain the role of these auxiliary fields.

The interactions of the scalar multiplet are determined by an arbitrary function V(Z) of the fields Z^a (with no $\bar{Z}^a$ dependence). This function is called the superspace potential, and we denote derivatives by

$$V_{,a} = \frac{\partial}{\partial Z^a}\ V(Z) \ ,$$

$$V_{,ab} = \frac{\partial^2}{\partial Z^a \partial Z^b} V(Z) \quad , \tag{38}$$

and introduce chiral projection matrices $L,R = (1 \mp \gamma_5)$. The interaction Lagrangian is then

$$L_{int} = \left[F^a\, V_{,a} + \overline{F}^a\, \overline{V}_{,a} + \overline{\psi}^a(LV_{,ab} + R\overline{V}_{,ab})\psi^b\right] \quad . \tag{39}$$

The auxiliary fields can be eliminated using their field equations, e.g., $\overline{F}^a = -V_{,a}$. Upon substitution of this in (37) and (39) one finds the physically equivalent Lagrangian

$$L = \partial_\mu \overline{Z}^a \partial_\mu Z^a + 1/2\; i\overline{\psi}^a \not\partial \psi^a - \overline{V}_{,a}\, V_{,a} + \overline{\psi}^a(LV_{,ab} + R\overline{V}_{,ab})\psi^b \quad . \tag{40}$$

The transformation rules of the scalar multiplet are

$$\begin{aligned} \delta Z^a &= \overline{\varepsilon} L \psi^a \quad , \\ \delta(L\psi^a) &= L(-i\not\partial Z^a + F^a)\varepsilon \quad , \\ \delta F^a &= -i\overline{\varepsilon}\not\partial L\psi^a \quad , \end{aligned} \tag{41}$$

with other variations determined by complex conjugation and charge conjugation.

The kinetic and interaction Lagrangians (37) and (39) are separately invariant under (41). After elimination of auxiliary fields, the Lagrangian (40) is invariant under the variations δZ^a and $\delta\psi^a$ of (41) with $F^a = -\overline{V}_{,a}$.

The renormalizability properties of the theory depend on the superpotential V(Z). If V(Z) is a cubic polynomial, viz.

$$V(Z) = b_a Z^a + m_{ab}\, Z^a Z^b + g_{abc}\, Z^a Z^b Z^c \quad ,$$

where m_{ab} and g_{abc} are mass and coupling matrices, then the Lagrangian (40) contains boson vertices up to quartic order plus Yukawa vertices. It is therefore renormalizable. If V(Z) contains terms of higher than cubic order, then the theory is non-renormalizable by power counting. In the renormalizable case, it is remarkable that compensation of divergences requires counterterms only with the structure of the kinetic Lagrangian (37). Another way to state this is that the only independent renormalization is a common wave-function renormalization of Z^a, ψ^a, F^a, and that renormalization of mass and coupling parameters is determined by wave-function renormalization. The best way to demonstrate these results is to use superspace perturbation theory.

Let us discuss auxiliary fields. The main point is that Z,ψ,F constitute a fundamental multiplet of fields on which the Poincaré supersymetry algebra acts as a closed algebra. Indeed, they are the components of a chiral superfield. What a closed algebra means is that the commutator of two supersymmetry transformations (41) with parameters ε_1 and ε_2 gives

$$[\delta_{\varepsilon_1}, \delta_{\varepsilon_2}] = i(\bar{\varepsilon}_1\gamma^\mu\varepsilon_2)\partial_\mu \quad (42)$$

on all fields, so that we obtain a translation in space-time as required by $\{Q,\bar{Q}\} = P$. If one evaluates this commutator after elimination of auxiliary fields using δZ and $\delta\psi$ of (41), then there are additional terms on the right side of (42) which vanish when the spinor equations of motion of the Lagrangian (40) are used, so that the algebra closes only on shell. It is only when auxiliary fields are used that the kinetic and interaction terms are separately invariant. More generally, one can develop a tensor calculus of multiplets and systematic procedures for constructing supersymmetric invariants. A more physical reason for the use of auxiliary fields is that they are usually necessary to characterize clearly the spontaneous breakdown of supersymmetry.

The auxiliary field structure of $N = 1$ supersymmetry and supergravity is fairly well understood and is closely related to the superspace formalism. There is also considerable information for $N = 2$ extended supersymmetry. However, auxiliary fields for $N \geqslant 2$ are an usolved problem to which a great deal of effort has been devoted.

Supergravity

We now continue this survey of $N = 1$ supersymmetric theories by discussing some of the motivations and features of "simple supergravity" in which supersymmetry holds as a local or gauge invariance, the spinor parameter $\varepsilon(x)$ is an arbitrary function of space-time coordinates. Certainly the major motivation for the development of supergravity is that the gauge form of a symmetry is more powerful than the global form. It was realized at an early stage (see the arguments below) that a locally supersymmetric theory necessarily includes gravity and that any supersymmetric theory with gravity must be locally supersymmetric. Since any theory of elementary particles must eventually include the gravitational force, there is compelling physical motivation for supergravity.

There is a simple argument that any theory of local supersymmetry must be a gravitational theory. It follows from the structure of the commutator of supersymmetry transformations in (42), that if supersymmetry is local, then the parameter of infinitesimal translations, $\bar{\varepsilon}_1(x)\gamma^\mu\varepsilon_2(x)$, will also be local. Thus one must have general coordinate invariance. The reason is that only covariant concepts can enter in a gravitational theory, whereas a constant spinor ε_α is not covariant since it transforms into a space-time dependent spinor under local Lorentz transformations,

$$\varepsilon \to \exp(1/2\ \Lambda_{ab}(x)\ \sigma^{ab})\varepsilon \quad . \tag{43}$$

One may reasonably speculate that the simplest supergravity theory corresponds to an irreducible representation of the algebra which contains the spin 2 graviton. The choices are the (2,3/2) and (5/2,2) representations. However, gauge fields in mathematical physics always have one vector index together with indices of the charge to be gauged. In supergravity, we wish to gauge a spinor charge Q_α and it is reasonable to suppose that the appropriate gauge field is the vector-spinor $\psi_{\rho\alpha}(x)$. This indicates that the particle content of supergravity is given by the (2,3/2) representation, since the vector-spinor field can describe spin 3/2, but not spin 5/2, particles.

The plausible line of reasoning above can succeed only if the spin 3/2 has a gauge invariance, a question which can be examined in the free-field limit. In 1941 Rarita and Schwinger proposed a Lagrangian which describes a massive spin 3/2 particle. This Lagrangian is equivalent to

$$L_{3/2} = 1/2\ \varepsilon^{\lambda\rho\mu\nu}(\bar{\psi}_\lambda\gamma_5\gamma_\mu\partial_\nu\psi_\rho) + 1/2\ m\bar{\psi}_\lambda\ \sigma^{\lambda\rho}\ \psi_\rho \quad . \tag{44}$$

For m = 0 the action is invariant under the gauge transformation

$$\delta\psi_\rho = \partial_\rho\varepsilon(x) \quad , \tag{45}$$

where $\varepsilon(x)$ is an arbitrary spinor function. Clearly this is the exact analog of the gauge transformation of electrodynamics, and it is exactly what is needed for supergravity. Note that there is a one-parameter continuum of equivalent spin 3/2 theories obtained by field redefinition

$$\psi'_\rho = \psi_\rho + A\gamma_\rho\gamma_\mu\psi^\mu \quad , \tag{46}$$

where A is a constant. The gauge transformation law is more complicated if such a field redefinition is made, and this explains why the form (44) is always used in the supergravity literature.

The massless spin 3/2 field must be quantized as a gauge field, for example, using a generalization of the Faddeev-Popov method. The result for the free field expansion is

$$\psi_\rho(x) = \sum_{\lambda=\pm} \int \frac{d^3p}{(2\pi)^{3/2}(2\omega)^{1/2}} \left\{b(\vec{p},\lambda)\varepsilon_\mu(\vec{p},\lambda)u(\vec{p},\lambda)e^{-ip\cdot x} + b^\dagger(\vec{p},\lambda)\varepsilon_\mu{}^*(\vec{p},\lambda)v(\vec{p},\lambda)e^{+ip\cdot x}\right\} \quad , \tag{47}$$

where transverse polarization vectors $\varepsilon_\mu(\vec{p},\lambda)$ and massless spinors $u(p,\lambda)$ appear. The total helicity of the wave functions of physical modes of the fields is ±3/2, which shows that the theory describes spin 3/2 quanta.

It is fairly straightforward to proceed from the present simple arguments to a fully interacting supergravity theory. However, some detailed knowledge of the vierbein formalism (which is necessary to describe spinors in general relativity) and the Palatini formalism (in which the gauge field of local Lorentz transformation is an independent field) is required. We therefore state the results briefly and refer readers to the recent Physics Report (Reference 6 below) for further details.

The fields of simple supergravity are the vierbein $V_{a\mu}(x)$, spin connection $\omega_{\mu ab}(x)$ and $\psi_{\rho\alpha}(x)$ which can be regarded as gauge potentials for the charges P_a, M_{ab}, and Q_α of the supersymmetry algebra. From these fields are formed the curvature tensor and local-Lorentz covariant derivative

$$\begin{aligned} R_{\mu\nu ab} &= \partial_\mu\omega_{\nu ab} - \partial_\nu\omega_{\mu ab} + \omega_{\mu a}{}^{c}\,\omega_{\nu cb} - \omega_{\nu a}{}^{c}\,\omega_{\mu cb} \quad , \\ D_\nu\psi_\rho &= (\partial_\nu + 1/2\;\omega_{\nu ab}\,\sigma^{ab})\psi_\rho \quad . \end{aligned} \tag{48}$$

The Lagrangian can then be expressed as a minimal coupling of these fields, namely

$$L = -\frac{\det V}{4\kappa^2}\; V^{a\mu}\, V^{b\nu}\, R_{\mu\nu ab} - 1/2\;\varepsilon^{\lambda\rho\mu\nu}\,\overline{\psi}_\lambda\gamma_5\gamma_\mu D_\nu\psi_\rho \quad , \tag{49}$$

where $\gamma_\mu = \gamma^a V_{a\mu}$ and $\kappa^2 = 4\pi G$ is the gravitational coupling constant. The field $\omega_{\mu ab}$ is non-propagating. Its field equations can be solved to give

$$\begin{aligned} \omega_{\mu ab} = 1/2\; &\big[V_a{}^{\nu}(\partial_\mu V_{b\nu} - \partial_\nu V_{b\mu}) + V_a{}^{\rho}V_b{}^{\sigma}(\partial_\sigma V_{c\rho})V^c{}_\mu \\ &+ i\kappa^2(\psi_\mu\gamma_a\psi_b - 1/2\;\psi_a\gamma_\mu\psi_b)\big] - \big[a \leftrightarrow b\big] \quad . \end{aligned} \tag{50}$$

The vierbein terms are those of the more standard second order formalism of relativity and are the vierbein analog of the standard expression of the Christoffel symbol $\Gamma^\rho{}_{\mu\nu}$ in terms of the metric $g_{\mu\nu}$. The term bilinear in ψ_μ can be described geometrically as torsion due to the spin 3/2 field.

If (50) is substituted in (49) one obtains the second order form of supergravity where the torsion terms are quartic in ψ_ρ. This second order action is invariant under the local supersymmetry transformation rules

$$\begin{aligned} \delta V_{a\mu} &= -i\kappa\;\overline{\varepsilon}(x)\gamma_a\psi_\mu \quad , \\ \delta\psi_\mu &= \kappa^{-1}\; D_\mu\varepsilon(x) \quad . \end{aligned} \tag{51}$$

The spin 3/2 transformation rule involves the covariant derivative of the gauge parameter as is the case for gauge potentials in Yang-Mills theory and ordinary gravitation. This shows that ψ_μ is indeed the gauge field of supersymmetry transformations.

One should note that the literature contains many previous attempts to introduce interactions for the Rarita-Schwinger field (and other types of fields with spin 3/2 quanta). Such theories were always plagued by inconsistency such as acausal propagation of signals. It is now known that the particular type of interactions which occur in supergravity are consistent because of the local supersymmetry which can be viewed as an extension of the free field gauge invariance (45) to include interactions. Without the supersymmetry algebra P_a, M_{ab}, and Q_α, it would have been difficult to guess that the correct generalization of this gauge invariance required gravitation.

Catalog of N = 1 Supersymmetric Theories

We present in Table I a listing of N = 1 supersymmetric theories, classified by representation of the algebra and field content with comments.

Table I. List of N=1 Supersymmetric Theories

	Fields	Remarks
(1/2, 0)	ψ, A, B	Mass m and renormalizable coupling g. Global Internal Symmetry.
(1, 1/2)	A^a, χ^a	Yang-Mills interactions if fields belong to adjoint rep. of gauge group. Otherwise a free theory.
(1, 1/2, 1/2, 0) $m \neq 0$	B_μ, ϕ real ψ complex	Supersymmetric extension of free massive vector field.
(3/2, 1)	ψ_μ, A_μ	Free theory.
(2, 3/2)	$V_{a\mu}$, ψ_μ	Simple supergravity.
(s, s−1/2) s≥5/2	$\phi_{\mu\nu\cdots}$ symmetric tensor $\psi_{\mu\nu\cdots}$ and tensor-spinors	Free theories with Fang-Fronsdal higher spin gauge fields and transformation rules of Curtright.

In addition to these theories for particles of a single irreducible representation of the supersymmetry algebra, there are various theories involving a combination of representations. For example, theories of (1/2, 0) and (1, 1/2) representations with local internal symmetry are the basis of recent research on supersymmetric grand unified theories. It is also known how to couple all the global supersymmetric theories listed above (for $s_{max} \leq 3/2$) to the (2, 3/2) supergravity gauge multiplet.

Extended Supersymmetry Algebras

Let us now return and refocus our attention on supersymmetry algebras rather than field theories, and let us recall that the Coleman-Mandula theorem seriously limits the possibilitites of Lie group symmetries in quantum field theory. Since the allowed mathematical framework has now been extended to graded Lie algebras, it is important to ask what is the most general graded Lie symmetry allowed in relativistic quantum theory? The answer to this question is contained in the theorems of Haag, Lopuszanski and Sohnius. The general line of their argument now follows.

If symmetry of a relativistic S-matrix is realized by a graded Lie algebra containing even elements B^a and odd elements F_α, then the Lie subalgebra of even elements is constrained by the Coleman-Mandula theorem to be a direct product of Poincaré and internal symmetry. The graded Jacobi identities can then be used to show that the odd elements must transform in (1/2, 0) and (0, 1/2) representations of the Lorentz group. They must be spinors.

Complete analysis of the graded Jacobi identities requires that the most general graded Lie algebra allowed has the following structure. The odd elements consist of N Majorana spinor charges Q^i with $i = 1,2,\ldots,N$, and their anticommutator is given by

$$\{Q_\alpha{}^i, \overline{Q}_\beta{}^j\} = \delta^{ij}(\gamma^\mu)_{\alpha\beta}\, P_\mu + i\, \delta_{\alpha\beta}\, U^{ij} + (\gamma_5)_{\alpha\beta}\, V^{ij} \quad . \quad (52)$$

The space-time translation occurs and in addition there may be Hermitian scalar and pseudo-scalar internal symmetry charges U^{ij} and V^{ij} which are antisymmetric in $i \leftrightarrow j$. These charges must be true central elements of the entire algebra, which means that they commute among themselves and with all other elements. The even elements consist of the Poincaré charges $M_{\mu\nu}$, P_μ, the central charges, and possible additional internal symmetry charges T^a (scalars) and $T_5{}^a$ (pseudo-scalars). These charges determine a Lie algebra, and spinor charges transform in a representation of that algebra specified by matrices $(t^a)^{ij}$ and $(s^a)^{ij}$, i.e.

$$\left[T^a, Q_\alpha{}^i\right] = -(t^a)^{ij}\, Q_\alpha{}^j \quad ,$$

$$[T_5^b, Q_\alpha{}^i] = -(\gamma_5)_{\alpha\beta} (s^b)^{ij} Q_\beta{}^j \quad . \tag{53}$$

Thus supersymmetry and internal symmetry are unified in a single algebraic system! These systems are called the N-extended Poincaré supersymmetry algebras where N is the total number of spinor charges.

Exactly which internal symmetry groups occur is left open by the general analysis and depends on the particular field theory under consideration. When central charges are absent, the typical symmetry group of the scalar charges T^a is SO(N), and this is usually enlarged to a chiral SU(N) or U(N) invariance by the pseudoscalar charges.

If there are central charges the internal symmetry group is usually smaller. Since the central charges have dimension 1, as is evident from (52), they can only occur in field theories where there is a dimensional parameter, and one can show further that central charges are non-trivial only in $m \neq 0$ representations of the full algebra. Although there is a rich structure associated with the central charges, we will ignore them henceforth in this introductory treatment.

It was implicitly assumed in the previous discussion that, as in the work of Coleman and Mandula, there are $m \neq 0$ single particle states in the theory under discussion. In a purely massless theory, there is typically additional symmetry known as conformal invariance. A supersymmetric extension of conformal symmetry was found by Wess and Zumino and is called the superconformal group, and its generalization to systems of N supersymmetry charges was considered by Haag, Lopuszanski and Sohnius. We will not consider conformal supersymmetry in any detail. However, one general statement which should be made is that any supersymmetric theory with ordinary conformal invariance automatically enjoys superconformal invariance. There are N additional spinor charges $S_\alpha{}^i$ generated by commutation of $Q_\alpha{}^i$ with conformal boosts K_μ, viz.

$$[K_\mu, Q_\alpha{}^i] = (\gamma_\mu)_{\alpha\beta} S_\beta{}^i \quad , \tag{54}$$

and a U(N) (SU(4) for N = 4) internal symmetry group which is unified with supersymmetry.

Representations of Extended Supersymmetry

We now derive the massless irreducible representations of supersymmetry. Specifically, we consider the algebraic system of N-spinor charges $Q_\alpha{}^i$, the Poincaré generators $M_{\mu\nu}$, P_μ and for simplicity, only scalar internal symmetry charges T^a. It will be clear how to include pseudoscalar charges. Further, the assumption that central charges vanish is no loss of generality in the massless case.

It is best to work in the γ-matrix representation (16-17) and decompose the Majorana spinor charges into irreducible two component spinors $Q_\alpha{}^i$ and their adjoints $Q_\alpha{}^{i*}$. The supersymmetry anticommutator (52) is then equivalent to the three statements

$$\{Q_\alpha{}^i, Q_\beta{}^{j*}\} = \delta^{ij}(\sigma^\mu)_{\alpha\beta} P_\mu \quad ,$$
$$\{Q_\alpha{}^i, Q_\beta{}^j\} = 0 \quad , \tag{55}$$
$$\{Q_\alpha{}^{i*}, Q_\beta{}^{j*}\} = 0 \quad ,$$

which involve the 2x2 Weyl matrices $\sigma^\mu = (I,\vec{\sigma})$. All structure relations of the algebra will be satisfied, but it is useful to display the even-odd commutators for angular momentum and internal symmetry:

$$\left[\vec{J}, Q_\alpha{}^i\right] = -1/2\ (\vec{\sigma})_{\alpha\beta}\, Q_\beta{}^i \quad , \tag{56}$$

$$\left[T^a, Q_\alpha{}^i\right] = -(t^a)^{ij}\, Q_\alpha{}^j \quad . \tag{57}$$

The irreducible representations will be obtained using Wigner's method of induced representations, that is, an extension of the "little group" method for the Poincaré group. Since the Poincaré generators are a subalgebra, we pick a basis $|\overline{P},\lambda\rangle$ of one-particle helicity states of momentum $\overline{P}_\mu = (\omega, o, o, \omega)$ in the z-direction. On this basis we look for a representation of the "little algebra" i.e., the subalgebra of Q, Q^*, $M^{\mu\nu}$, T^a which leaves this basis invariant. The little algebra consists of the operators Q, Q^*, and T^a which commute quite generally with P, and the operators J_3, $M^{01}-J_2$, $M^{02}+J_1$ which generate the well known E_2 subgroup of the Poincaré group and have standard action on helicity states.

In the basis $|\overline{P},\lambda\rangle$ the anticommutators (55) become

$$\{Q_2{}^i,Q_2{}^{j*}\} = 2\omega\, \delta^{ij} \quad ,$$
$$\{Q_1{}^i,Q_1{}^{j*}\} = 0 \quad ,$$
$$\{Q_\alpha{}^i,Q_\beta{}^j\} = 0 \quad , \tag{58}$$
$$\{Q_\alpha{}^{i*},Q_\beta{}^{j*}\} = 0 \quad ,$$

where $\{Q_1,Q_1{}^*\}$ vanishes due to a cancellation between the 0 and 3 components in $\sigma^\mu\overline{P}_\mu$. Positivity then requires that the operators $Q_1{}^i$ and $Q_1{}^{i*}$ be represented trivially, i.e., by the zero matrix. To see this, note that for fixed i = j, one has the structure

$$MM^* + M^*M = 0 \quad . \tag{59}$$

Now take expectation values and sum over intermediate states to obtain

$$\sum_i \left(|\langle i|M|j\rangle|^2 + |\langle j|M|i\rangle|^2 \right) = 0 \quad , \tag{60}$$

which implies $\langle i|M|j\rangle = 0$.

The algebra of the remaining operators $Q_2{}^i$ and $Q_2{}^{i*}$ reduces to the anti-commutation rules of N independent fermion annihilators and creators, that is, a Clifford algebra of N complex elements. Further from (56) one finds that $Q_2{}^i$ raises and $Q_2{}^{i*}$ lowers helicity by 1/2 unit. It is now very easy to find the irreducible representations using Fock space techniques.

One chooses the "Clifford vacuum" $|\overline{P},\overline{\lambda}\rangle$ to have the properties

$$Q_2{}^i|\overline{P},\overline{\lambda}\rangle = Q_1{}^i|\overline{P},\overline{\lambda}\rangle = Q_1{}^{i*}|\overline{P},\overline{\lambda}\rangle = 0 \quad . \tag{61}$$

This implies that $Q_2{}^i$ are annhilators and $Q_2{}^{i*}$ are creators and that $\overline{\lambda}$ can be interpreted as the maximum helicity in the representation. The other relations in (61) merely restate the fact that the operators Q_1 and $Q_1{}^*$ are trivial. Temporarily we also assume that the Clifford vacuum is a singlet of the internal symmetry, i.e., $T^a|\overline{P},\overline{\lambda}\rangle = 0$.

By applying creators successively, one finds the sequence of states of indicated quantum numbers and multiplicity

$\vert\overline{P},\overline{\lambda}\rangle$	1 state
$\vert\overline{P},\overline{\lambda}-1/2,i\rangle = Q_2{}^{i*}\vert\overline{P},\overline{\lambda}\rangle$	N states
$\vert\overline{P},\overline{\lambda}-1,[ij]\rangle = Q_2{}^{j*}Q_2{}^{i*}\vert\overline{P},\overline{\lambda}\rangle$	1/2 N(N-1) states
.	
.	
.	
$\vert\overline{P},\overline{\lambda}-1/2\ N\rangle = Q_2{}^{N*}Q_2{}^{(N-1)*}\cdots Q_2{}^{1*}\vert\overline{P},\overline{\lambda}\rangle$	1 state

States of helicity $\overline{\lambda}-1/2$ m have multiplicity $N!/(m!(N-m)!)$ and the sequence stops when the singlet state of helicity $\overline{\lambda}-1/2$ N is reached. The total dimension of the representation is 2^N as is required by the representation theory of Clifford algebras. There are always an equal number of boson and fermion states.

As far as internal symmetry is concerned, the states of helicity $\overline{\lambda}$ - 1/2 m transform as m-fold antisymmetric products of the representation $(t^a)^{ij}$, i.e., as m-th rank totally antisymmetric tensors in the usual case when the $Q_\alpha{}^i$ transform in the vector representation of SO(N). One can also consider multiple Clifford vacua $|\bar{P},\bar{\lambda},t\rangle$ which transform in some irreducible representation of internal symmetry of dimension M. In this case, the total number of states obtained by applying creators $Q_2{}^{i*}$ is $(2^N)M$ and Clebsch-Gordan decomposition is necessary to describe the internal symmetry properties in detail.

Thus an irreducible representation is a "tower" of helicity states which contain particles of maximum helicity $\bar{\lambda}$ and minimum helicity $\bar{\lambda}$ - 1/2 N. In any local field theory, particle states occur in CPT conjugate pairs of helicity $\pm\lambda$. Therefore, a supersymmetric field theory usually describes a reducible representation of the algebra in which one adjoins the CPT conjugate states of reversed helicity to the initial sequence. Those states are obtained by starting from the CPT conjugate Clifford vacuum $|\bar{P},-\bar{\lambda}\rangle$ and applying the helicity raising operators $Q_2{}^i$. If the initial sequence is symmetric about zero, i.e., if $\bar{\lambda} - 1/2\, N = -\bar{\lambda}$, then it is CPT self-conjugate and there is no need to double the representation. This means that field theories for $N = 4\bar{\lambda}$ and $N = 4\bar{\lambda} - 1$ always have the same particle content. These self-conjugate theories seem to have special properties, as we will discuss.

To derive the $m \neq 0$ representations of extended supersymmetry, one proceeds in an analogous way, introducing a basis of helicity states in the rest frame $\bar{P}^\mu = (M,0,0,0)$. One finds that both $Q_1{}^i$ and $Q_2{}^i$ are non-trivial Clifford elements, and the representations are larger, containing 2^{2N} states. There are special mechanisms which lead to smaller $m \neq 0$ representations when there are central charges.

One important point is that the number of distinct spins increases with N in any representation. The lowest spin representation for given N is a tower of spins $s = 0,1/2,1\ldots,s_{max}$. For massless representations s_{max} = 1/4 (N+1) or 1/4 N for odd and even N respectively. For the massive case s_{max} = 1/2 N in representations without central charges. Clearly the possible interacting quantum field theories are limited by this spin explosion.

The field theories of extended supergravity can be divided in two classes:

i) theories with global supersymmetry and s_{max} = 1/2 or 1 exist for N = 1,...,4.

ii) supergravity theories with s_{max} = 2 exists for N = 1,...,8.

N = 4 Supersymmetric Yang-Mills Theory

This is the field theory of the self-conjugate representation of extended supersymmetry with N = 4 and $\overline{\lambda} = 1$. It is the maximal global supersymmetric theory and most interesting theoretically. All fields belong to the adjoint representation of an arbitrary gauge group G, and there is a global SU(4) internal symmetry, with a central SO(4) subgroup of scalar charges (under which the $Q_\alpha{}^i$ transform in the (1/2,1/2) representation). The notation and classification of fields is given in Table II.

The Lagrangian of the theory is

$$L = -1/4(F_{\mu\nu}{}^a)^2 + 1/2\, i\, \overline{\chi}^{ai} \gamma^\mu (D_\mu \chi)^{ai} + 1/2\, (D_\mu A^{aI})^2$$
$$+ 1/2\, (D_\mu B^{aI})^2 + g\, f^{abc}\, \overline{\chi}^{aj} (\alpha^I{}_{jk}\, A^{bI} + i\, \gamma_5 \beta^I{}_{jk} B^{bI}) \chi^{ck}$$
$$+ 1/4\, g^2 \{ (f^{abc}\, A^{bI}\, A^{cJ})^2 + (f^{abc}\, B^{bI}\, B^{cJ})^2$$
$$+ 2\, (f^{abc}\, A^{bI}\, B^{cJ})^2 \} \quad . \qquad (62)$$

All derivatives are gauge covariant derivatives with f^{abc} the structure constants of G. The 4x4 matrices $\alpha^I{}_{jk}$ and $\beta^I{}_{jk}$ are coupling matrices of the (1,0) or (0,1) and the (1/2,1/2) irreps. of SO(4). Therefore, they coincide with the η and $\overline{\eta}$ matrices of instanton theory.

One sees that the interactions are of minimal coupling Yang-Mills type together with Yukawa and ϕ^4 interactions determined by the same gauge coupling constant g. Thus, a priori, we expect a renormalizable theory with infinite renormalization of the coupling constant in every order of perturbation theory. It is very surprising that explicit Feynman graph calculations up to 3-loop order reveal that there is no charge renormalization, and the function $\beta(g)$ vanishes. One can also perform calculations in a gauge for superspace perturbation theory where all other divergences cancel. Most physicists believe that this theory will be ultraviolet finite

Table II. Fields for N=4 Supersymmetric Yang-Mills Theory

Spin	Multiplicity	Fields	SO(4) irrep.
1	1	$A_\mu{}^a$	(0,0)
1/2	4	χ^{ai}, i=1,...4	(1/2,1/2)
$0^\pm$	3 + 3	A^{aI}, B^{aI}, I = 1,2,3	(1,0) and (0,1)

to all orders, but a convincing proof of this remarkable property has never been given. One hope is that ultraviolet finiteness will be easy to prove, when and if an "off-shell formulation" of the theory with a full set of auxiliary fields for N = 4 supersymmetry is obtained.

Let us give the supersymmetry transformation rules of this theory as an example of a theory with extended supersymmetry. There are four transformation parameters ε^i corresponding to the four $Q_\alpha{}^i$, and variations take the form:

$$\begin{aligned}
\delta A_\mu &= i\,\bar{\varepsilon}^i \gamma_\mu \chi^i \quad , \\
\delta \chi^i &= \sigma^{\mu\nu} F_{\mu\nu}\, \varepsilon^i + (\not{\partial} A^I\, \alpha^I{}_{ij} + i\,\gamma_5 \not{\partial} B^I\, \beta^I{}_{ij})\varepsilon^j + \ldots \quad , \\
\delta A^I &= \bar{\varepsilon}^i\, \alpha^I{}_{ij}\, \chi^j \quad , \\
\delta B^I &= \bar{\varepsilon}^i \gamma_5\, \beta^I{}_{ij}\, \chi^j \quad ,
\end{aligned} \tag{63}$$

which combine terms previously found for N = 1 vector and scalar multiplets ((25) and (41)). Indeed, the N = 4 theory can be expressed as an N = 1 theory of one vector multiplet and three scalar multiplets, and the omitted terms in (63) correspond to the auxiliary fields in the scalar multiplet transformation rules.

What is the significance of this theory with its remarkable ultraviolet behavior? The model is probably not useful for phenomenology for several reasons, most notably that supersymmetry is exact, rather than spontaneously broken. It remains as a unique four-dimensional example of a field theory where ultraviolet behavior is better than expected from standard power counting and supports hope for analogous miraculous finiteness in supergravity.

Extended Supergravity

Extended supergravity theories are field theories of $\lambda = 2$ representations of the supersymmetry algebra which exist for $1 \leq N \leq 8$ and which have manifest SO(N) internal symmetry. The particle content of these theories is given in Table III (where the entry denotes the dimensionality of the SO(N) representation for each spin).

The most remarkable property is the unification of particles of different spin. Starting from the spin 2 graviton, one can reach all of the particle states of the theory by combined supersymmetry and SO(N) transformations. All particles are unified with the graviton.

In principle, the supergravity fields theories can unify gravitation with Yang-Mills interactions and contain fundamental spin 1/2 and spin 0 fields in a very rigid framework. Unfortunately, it appears that the theories are too small to be used directly for grand

Table III. Particle content of N-extended Supergravity

	N = 1	N = 2	N = 3	N = 4	N = 5	N = 6	N = 7	N = 8
S=2	1	1	1	1	1	1	1	1
S=3/2	1	2	3	4	5	6	7+1	8
S=1		1	3	6	10	15+1	21+7	28
S=1/2			1	4	10+1	20+6	35+21	56
S=0				1+1	5+5	15+15	35+35	35+35

unified phenomenology. The basic reason is that the SO(8) gauge group is too small to contain the phenomenological group SU(3)xSU(2)xU(1), and the quantum members of the 28 spin 1 and 56 spin 1/2 states are in imperfect correspondence with present "elementary" particles. Since the natural energy scale of gravitation is 10^{19} GeV, we may hope that supergravity gives a correct preon theory with its fundamental degrees of freedom detectable only at extremely high energy, while present quarks, leptons, and vector bosons are composite states.

The Lagrangians of extended supergravity are relatively simple for $N \leq 3$ and quite complicated for $N \geq 4$ when scalar fields enter. For illustration, we give the N = 3 theory with vierbein $V_{a\mu}$, triplets of gravitino and vector fields $\psi_\mu{}^i$ and $A_\mu{}^i$ and singlet spinor χ. The Lagrangian is

$$\begin{aligned} L = &-\frac{V}{4\kappa^2} V^{a\mu} V^{b\nu} R_{\mu\nu ab} - 1/2\, \varepsilon^{\lambda\rho\mu\nu} \bar{\psi}_\lambda{}^i \gamma_5 \gamma_\mu (D_\nu \psi_\rho)^i \\ &- 1/4\, V (F_{\mu\nu}{}^i)^2 + 1/2\, i\, V \bar{\chi} \not{\partial} \chi - 1/2\, \kappa \varepsilon^{ijk} \bar{\psi}_\mu{}^i (V F^{\mu\nu k} \\ &- 1/2\, i \gamma_5 \tilde{F}^{\mu\nu k}) \psi_\nu{}^j - \frac{i}{\sqrt{2}} \kappa V \bar{\psi}_\lambda{}^i \sigma^{\mu\nu} F_{\mu\nu}{}^i \gamma^\lambda \chi \\ &+ \text{quartic fermion terms} \quad , \end{aligned} \tag{64}$$

where $V = \det V_{a\mu}$, $F_{\mu\nu}{}^i = \partial_\mu A_\nu{}^i - \partial_\nu A_\mu{}^i$ and $\tilde{F}_{\mu\nu}{}^i$ is its dual.

Although the scalar interactions in $N \geq 4$ supergravity are non-polynomial and appear complicated, there is an important general feature. Namely, the internal space of the scalar field variables has the geometry of a non-compact quotient space, and the scalar Lagrangian has the form of a non-linear σ-model. In the N = 4 model the internal space is SU(1,1)/U(1), and the scalar kinetic Lagrangian for the fields A(x) and B(x) is

$$L_{scalar} = \left[2(1-\kappa^2(A^2+B^2))\right]^2 \left((\partial_\mu A)^2 + (\partial_\mu B)^2\right) \quad . \qquad (65)$$

For N = 8 the internal space is E(7)/SU(8), and the Lagrangian is written with explicit nonlinear E(7) invariance, and an SU(8) gauge invariance with 56 auxiliary vector potentials without conventional Yang-Mills kinetic terms. Thus the theory has unexpected symmetries. One speculates that propagating SU(8) gauge bosons are generated dynamically and that this leads to an effective grand unified theory with SU(8) gauge group.

The coupling constants of extended supergravity theories are Newton's constant $\kappa^2 = 4\pi G$, and a gauge coupling constant e of local SO(N) invariance. The properties of the theories discussed in the last few paragraphs assumed e = 0 and global SO(N) invariance. It is known that the gauging of SO(N) is linked with large cosmological terms in the Lagrangian for $N \leqslant 3$ and to scalar field potentials which are unbounded from below for $N \geqslant 4$. Thus there are very puzzling open problems connected with the vacuum energy in this form of supergravity. The gauged form of the N = 8 supergravity has been obtained very recently. The SO(N) gauging is compatible with local SU(8) invariance, but breaks E(7) symmetry. This work appears to bring an era of work in supergravity to a close. The Lagrangians and transformation rules of all eight extended supergravities are now known in (essentially) their most general form. My view is that one should now seriously try to understand the vacuum energy problem of this form of supergravity.

Conclusions

I hope that I have conveyed in these lectures a picture of supersymmetry as a symmetry which operates in quantum field theory in a rather conventional way, but which has the unique power to unify particles of different spin at the global level and to unify gravitation with other forces at the local level. The development here has been restricted to one viewpoint, and readers will find it very useful to study the contributions of other lecturers at this school and previous pedagogical material on supersymmetry which contain many references to the research literature.

References

1. P. Fayet and S. Ferrara, "Supersymmetry", Physics Reports 32C (1977) 350.
2. Recent Developments in Gravitation, Cargese 1978, Maurice Levy and S. Deser, Editors, NATO Advanced Institute Series B, Vol. 44, Plenum Press, New York 1979.

3. Supergravity, P. van Nieuwenhuizen and D. Z. Freedman, Editors, North-Holland Publishing Co., Amsterdam, 1980.
4. Unification of the Fundamental Particle Interactions, Sergio Ferrara, John Ellis, and Peter van Nieuwenhuizen, Editors, Ettore Majorana International Science Series, Physical Sciences Vol. 7, Plenum Press, New York, 1980.
5. Superspace and Supergravity, Proceedings of the Nuffield Workshop, S.W. Hawking and M. Rocek, Editors; Cambridge University Press, Cambridge, 1981.
6. P. van Nieuwenhuizen, "Supergravity", Physics Reports Vol. 68, Number 4, February 1981.

SUPERFIELDS

M. T. Grisaru*

California Institute of Technology

Pasadena California 91125 USA

Introduction

This is an introduction to the use of superfields and superspace methods in supersymmetry. Superfields were introduced by Salam and Strathdee,[1] shortly after the initial work by Wess and Zumino.[2] They have proven to be a useful way of describing globally supersymmetric models, and superspace methods exist and are being refined for handling theories with local supersymmetry. In the past few years the small community of supergravity practitioners has split up somewhat into component people and superfield people with the former pointing out to the latter that practically all the results in supergravity were first discovered by component (ordinary field theory) methods. This is almost true (I believe W. Siegel may have anticipated some of the component results). Undoubtedly we are more at home with ordinary space field theory and our physical intuition and mathematical skills are better developed there. But superfields are an extremely useful tool for describing the properties of known supersymmetric systems (at least the ones we know completely, i.e. including auxiliary fields) and I suspect the day may come when we look at old papers on component supergravity with the same mixture of horror and admiration with which I look at say, Whittaker's "Analytical Dynamics" where he writes down and solves component by component what I recognize as vector equations.

My approach in these lectures is to concentrate primarily on superspace and superfields and only occasionally make contact with

*On leave from Brandeis University, Waltham MA 02254

component results. People who are familiar with ordinary methods will recognize and be familiar with these points of contact. To others I apologize and offer the excuse of time and space limitations. The same excuse is an explanation for the fact that many of the results will be presented without a derivation. I hope to present an overview of the subject, with enough details so that with some work all the skipped steps can be filled in and so that the available literature will be more accessible.

References and notation

Superfields are discussed in a long, old article by Salam and Strathdee,[3] and in the Fayet and Ferrara review article.[4] Geometrical aspects of superspace and supergravity are discussed in Cargese lectures by Zumino,[5] Salamanca lectures by Wess,[6] and Trieste lectures by Grimm.[7] An up to date description of superspace and superfields, including the Siegel-Gates superfield description of supergravity,[8] can be found in the Cambridge lectures by Rocek.[9] Superfield perturbation theory is also reviewed in this reference and in my Trieste lectures.[10] All of these lectures contain references to the original articles. While superspace methods in global supersymmetry are quite elementary and easily learned, I have not found this to be the case for local supersymmetry. Whatever understanding I have of the subject comes primarily from Zumino's Cargese lectures, and very rewarding if not always easy forays into the various Gates and Siegel papers (Nuclear Physics and Physics Letters from about 1978 on).

The question of notation and conventions is always somewhat vexing. Most superspace practitioners are converging towards the use of two component (Weyl) spinors, whereas the original formulation and most of the early work used four component spinors. Where differences appear are in the choice of factors of 2, i, metric and summation conventions. My conventions are the same as Rocek's,[9] but differ in some details from those of Zumino and others. Hence some peculiar minus signs and factor differences.

We use two component anticommuting spinors with dotted and undotted indices and the correspondence

$$1/2(1+\gamma_5)\psi \leftrightarrow \psi^{\alpha} , \quad 1/2(1-\gamma_5)\psi \leftrightarrow \bar{\psi}_{\dot{\alpha}} .$$

We raise and lower indices with

$$\varepsilon_{\alpha\beta} = -\varepsilon_{\beta\alpha} , \quad \varepsilon^{\alpha\beta}\,\varepsilon_{\gamma\beta} = \delta^{\alpha}{}_{\gamma} ,$$

$$\psi_{\alpha} = \psi^{\beta}\,\varepsilon_{\beta\alpha} , \quad \psi^{\alpha} = \varepsilon^{\alpha\beta}\,\psi_{\beta} ,$$

with exactly the same conventions for dotted indices (this differs from others' conventions). We define

$$\psi^2 = \psi^\alpha\psi_\alpha \ , \quad \bar\psi^2 = \bar\psi^{\dot\alpha}\bar\psi_{\dot\alpha} \ ,$$

$$\psi^\alpha\psi_\beta = 1/2 \ \delta^\alpha{}_\beta\psi^2 \ .$$

Note that $(\psi)^3 = 0$. Tensor indices are related to spinor indices by

$$q_{\alpha\dot\alpha} = \sigma^a{}_{\alpha\dot\alpha} \, q_a \ , \quad q_a = 1/2 \ \sigma_a{}^{\alpha\dot\alpha} \, q_{\alpha\dot\alpha} \ ,$$

$$\sigma^a = (1,\vec\sigma) \ , \quad \sigma^a{}_{\alpha\dot\alpha} \, \sigma_a{}^{\beta\dot\beta} = 2\delta^\beta{}_\alpha \, \delta^{\dot\beta}{}_{\dot\alpha} \ , \quad \sigma^a{}_{\alpha\dot\alpha} \, \sigma_b{}^{\alpha\dot\alpha} = 2\delta^a{}_b \ .$$

Superspace will be parametrized by space-time coordinates x^a and anticommuting spinor coordinates θ^α, $\bar\theta^{\dot\alpha}$. Spinorial derivatives are defined by

$$\partial_\alpha\theta^\beta = \delta^\beta{}_\alpha \ , \quad \partial_{\dot\alpha}\bar\theta^{\dot\beta} = \delta^{\dot\beta}{}_{\dot\alpha} \ .$$

and integration over spinor variables is defined such that

$$\int d\theta^\alpha \, \theta_\beta = -i/2 \ \delta^\alpha{}_\beta \ ,$$

while all other integrals are zero. This particular normalization is used here for convenience. Thus for $f(\theta) = f_0 + \theta^\alpha f_{1\alpha} + \ldots$

$$\int d\theta^\alpha \, f(\theta) = i/2 \ \partial^\alpha f \, |_{\theta=0} \ .$$

We note also the double integral

$$\int d^2\theta \ \theta^2 = \int d\theta^\alpha \, d\theta_\alpha \, \theta^\beta \, \theta_\beta = 1 \ ,$$

which can be written symbolically

$$\int d^2\theta = -1/4 \ \partial^\alpha\partial_\alpha \ .$$

Finally we shall denote integration over all four θ's by $\int d^4\theta$.

Superfields

The global supersymmetry algebra

$$\{Q_\alpha,\bar Q_{\dot\alpha}\} = -2i \ \sigma^m{}_{\alpha\dot\alpha} \, P_m \ ,$$

$$\{Q_\alpha,Q_\beta\} = \{\bar Q_{\dot\alpha},\bar Q_{\dot\beta}\} = [Q_\alpha,P_m] = 0 \ ,$$

$$[Q_\alpha,M_{mn}] = (\sigma_{mn})_{\alpha\beta} \, Q^\beta, \ \text{etc.}$$

can be realized linearly as the algebra of translations and Poincaré transformations in superspace $(x^a,\theta^\alpha,\bar\theta^{\dot\alpha})$ by operators acting on superfields

$$\Psi_{\gamma\delta\ldots,\dot{\gamma}\dot{\delta}\ldots}(x^a,\theta^\alpha,\bar{\theta}^{\dot{\alpha}}) \quad .$$

Here $\gamma\delta\ldots,\dot{\gamma}\dot{\delta}\ldots$ denote Lorentz spinor (or equivalently tensor) indices. Elements of the Poincaré algebra act in standard fashion on the Ψ's (the θ's transforming under SL(2c)), $P_m = i\frac{\partial}{\partial x^m} = i/2\ \sigma_m{}^{\alpha\dot{\alpha}}\ \partial_{\alpha\dot{\alpha}}$, etc., while the Q's are represented by

$$Q_\alpha = \frac{\partial}{\partial\theta^\alpha} - i\bar{\theta}^{\dot{\alpha}}\ \partial_{\alpha\dot{\alpha}} \quad ,$$

$$\bar{Q}_{\dot{\alpha}} = \frac{\partial}{\partial\bar{\theta}^{\dot{\alpha}}} - i\theta^\alpha\ \partial_{\alpha\dot{\alpha}} \quad .$$

Thus they act as generators of superspace translations:

$$e^{i(\varepsilon^\beta Q_\beta + \bar{\varepsilon}^{\dot{\beta}} \bar{Q}_{\dot{\beta}})}\ \Psi(x^m,\theta^\alpha,\bar{\theta}^{\dot{\alpha}}) = \Psi(x^m - \varepsilon\sigma^m\bar{\theta} - \theta\sigma^m\bar{\varepsilon},\theta + \varepsilon,\bar{\theta} + \bar{\varepsilon})$$

and it is easy to check that indeed the algebra is satisfied.

Superfields can be expanded in a Taylor series in $\theta,\bar{\theta}$:

$$\Psi(x,\theta,\bar{\theta}) = C(x) + \theta^\alpha\chi_\alpha(x) + \bar{\theta}^{\dot{\alpha}}\bar{\chi}_{\dot{\alpha}}(x) + \theta^2 M(x) + \bar{\theta}^2 N(x)$$
$$+ \theta\sigma_a\bar{\theta}\ A^a(x) + \bar{\theta}^2\theta^\alpha\psi_\alpha(x) + \theta^2\bar{\theta}^{\dot{\alpha}}\bar{\psi}_{\dot{\alpha}}(x) + \theta^2\bar{\theta}^2 D(x) \quad .$$

The expansion stops at the $\theta^2\bar{\theta}^2$ level because $(\theta)^3 = (\bar{\theta})^3 = 0$. The quantities $C(x)$, $\chi(x)$,... are the component fields of Ψ. They also carry the explicit spinor or tensor indices that Ψ may carry. The supersymmetry transformation

$$\delta_\varepsilon\Psi = (\varepsilon^\alpha\ Q_\alpha + \bar{\varepsilon}^{\dot{\alpha}}\ \bar{Q}_{\dot{\alpha}})\Psi$$

induces transformations among the components of Ψ which are the usual transformations of component supersymmetry. For example, writing

$$\delta\left[C + \theta^\alpha\chi_\alpha + \ldots\right] = \left[C(x^m - \varepsilon\sigma^m\bar{\theta} - \theta\sigma^m\bar{\varepsilon}) + \ldots\right] - \left[C(x) + \ldots\right]$$

and rearranging the right hand side in powers of $\theta,\bar{\theta}$ gives the usual supersymmetry transformations of component supersymmetry:

$$\delta C(x) = \varepsilon^\alpha\chi_\alpha \ , \qquad \delta\chi(x) = 2M\varepsilon - \sigma^m\partial_m C\bar{\varepsilon},\ldots \quad ,$$

etc. In particular one discovers that the variation of the last component of the superfield is a space-time derivative. That is why supersymmetric component actions can be constructed out of the D-components of (suitable products of) superfields of this type.

The θ's are assigned (mass) dimension -1/2 so that the component fields have increasing dimension from dimC to dimD=dimC+2. It is clear therefore that not all the components can be identified with physical propagating fields which should have dimensions 1 or 3/2 to allow writing of ordinary kinetic terms e.g. $A \Box A$ or $\psi^\alpha \partial_{\alpha\dot{\alpha}} \overline{\psi}^{\dot{\alpha}}$. The higher dimension component fields generally appear in the Lagrangian without derivatives and can therefore be eliminated by using their (algebraic) equations of motion; they are the "auxiliary fields". The lower dimension component fields are often gauge components and do not appear in the Lagrangian. Finally we make the trivial observation that the component fields can be defined as θ-derivatives of Ψ at $\theta,\overline{\theta} = 0$.

While space-time derivatives of superfields are also superfields ($i\frac{\partial}{\partial x^m} = P_m$ commutes with the Q's) this is not the case with the spinorial derivatives ∂_α. However it is easy to find covariant derivatives which do. They are given by

$$D_\alpha = i/2(\partial_\alpha + i\overline{\theta}^{\dot{\alpha}} \partial_{\alpha\dot{\alpha}}) \quad ,$$

$$\overline{D}_{\dot{\alpha}} = -i/2(\partial_{\dot{\alpha}} + i\theta^\alpha \partial_{\alpha\dot{\alpha}}) \quad ,$$

satifiying the anticommutation relations

$$\{D_\alpha, \overline{D}_{\dot{\alpha}}\} = i/2 \; \partial_{\alpha\dot{\alpha}} \quad ,$$

$$\{D_\alpha, D_\alpha\} = \{\overline{D}_{\dot{\alpha}}, \overline{D}_{\dot{\alpha}}\} = 0 \quad ,$$

hence, in particular $(D)^3 = (\overline{D})^3 = 0$. It is easy to verify the D's and the Q's anticommute so that if Ψ is a superfield so is $D_\alpha\Psi$ and $\overline{D}_{\dot{\alpha}}\Psi$. For future use we note that the following quantities

$$\Pi_{1/2} = -2\,\frac{D^\alpha \overline{D}^2 D_\alpha}{\Box} \; , \; \Pi_{0+} = \frac{\overline{D}^2 D^2}{\Box} \; , \; \Pi_{0-} = \frac{D^2 \overline{D}^2}{\Box} \quad ,$$

are projection operators and satisfy $\Pi_{1/2} + \Pi_{0+} + \Pi_{0-} = 1$. They are often useful for decomposing superfields into irreducible (under supersymmetry) parts. Finally we observe that the D's obey the Leibnitz rule and behave in all respects like ordinary derivatives (except for anticommutativity).

Irreducible representations and chiral superfields

A glance at the expansion of the general superfield Ψ into components might suggest that it does not carry an irreducible representation of superysmmetry. There are several boson and fermion fields present, and this indicates that perhaps Ψ could be decomposed into several irreducible components. There exist general methods for constructing projection operators onto irreducible representations using Casimir operators,[11] or decomposition into "chiral" components,[12] but in the simplest cases to be discussed here one can proceed by direct construction. For simplicity consider the case of scalar superfields (i.e. no explicit Lorentz indices). One way to reduce a representation is to impose an invariant constraint on the superfield. A possible algebraic one turns out to be a reality condition. A superfield $V(x,\theta,\bar{\theta}) = V^*(x,\theta,\bar{\theta})$ will be called real and, as it turns out, is the suitable object needed to describe the (1/2,1) vector multiplet.

Turning to differential constraints, we are familiar with the constraint $\partial_m A^m = 0$ which separates out the spin 1 representation contained in a vector field, and by analogy we can try one of the covariant constraints $\bar{D}\Psi = 0$, $D\Psi = 0$, $D^2\Psi = 0$, or $\bar{D}^2\Psi = 0$. These turn out to be the only nontrivial ones. Since the D's commute with the generators of the algebra these are suitable constraints and, except for a possible reality condition completely reduce the general scalar superfield. An elementary although tedious way to see this is to first observe that the projection operators introduced above, when applied to a general supefield Ψ decompose it into pieces which satisfy the above constraints. A brute force calculation would then show that the various pieces contain just one independent fermion field. Since we expect any nontrivial representation of the algebra to contain at least one fermion field we should not expect to be able to reduce further.

A (complex) superfield $\Phi(x,\theta,\bar{\theta})$ satisfying the constraint $\bar{D}_{\dot{\alpha}}\Phi=0$ is called chiral. A superfield $\bar{\Phi}$ satisfying $D_\alpha\bar{\Phi} = 0$ is called antichiral. They and the real superfield V turn out to be the most useful ones for describing globally supersymmetric models. Linear superfields, satisfying $D^2L = 0$ or $\bar{D}^2L = 0$ often appear as gauge parameters and (gauge) field strengths but do not seem suitable for describing the fundamental quantum fields of a supersymmetric theory.

The real scalar superfield V and, in the case of supergravity, the real axial-vector superfield $H^a(x,\theta,\bar{\theta})$ turn out to describe gauge systems while the chiral and antichiral ones generally describe "matter" (spin 0 and spin 1/2) systems. (However, for example, a chiral spinor superfield Φ_α, $D\Phi_\alpha = 0$ does have gauge degrees of freedom). Let us discuss chiral superfields in some detail.

The constraint $\bar{D}_{\dot{\alpha}}\Phi(x,\theta,\bar{\theta}) = 0$ can be viewed as a first order differential equation. Its solution is simply

$$\Phi(x,\theta,\bar{\theta}) = e^{i\theta^\alpha\bar{\theta}^{\dot{\alpha}}\partial_{\alpha\dot{\alpha}}}\phi(x,\theta) \quad ,$$

where

$$e^{i\theta^{\alpha}\bar{\theta}^{\dot{\alpha}}\partial_{\alpha\dot{\alpha}}} = 1 + i\theta^{\alpha}\bar{\theta}^{\dot{\alpha}}\partial_{\alpha\dot{\alpha}} + 1/4\ \theta^2\bar{\theta}^2\Box \quad ,$$

while

$$\phi(x,\theta) = a + \theta^{\alpha}\psi_{\alpha} + \theta^2 f \quad ,$$

$$a = \frac{A+iB}{\sqrt{2}} \ , \ f = \frac{F-iG}{\sqrt{2}} \quad ,$$

is an arbitrary object depending on x and θ (not $\bar{\theta}$). A,B,Ψ may be identified with the physical scalar, pseudoscalar and spinor component fields of the Wess-Zumino scalar multiplet while F,G are the auxiliary fields. Their role, besides facilitating the writing of invariant component actions, is to allow the definition on the component fields of supersymmetry transformations which satisfy a closed algebra. In superspace of course they are an inseparable part of the superfields.

Similarly the antichiral constraint $D_{\alpha}\bar{\Phi}(x,\theta,\bar{\theta}) = 0$ gives

$$\bar{\Phi} = e^{-i\theta^{\alpha}\bar{\theta}^{\dot{\alpha}}\partial_{\alpha\dot{\alpha}}}\ \phi^*(x,\bar{\theta}) \quad .$$

(We have used a notation to indicate that ϕ^* is the complex conjugate of ϕ, which may be the case, but in general chiral and antichiral fields need not have anything to do with each other). We see in the above the explicit reduction of the general superfield to a smaller number of independent components. Of course the θ-expansion of Φ or $\bar{\Phi}$ has just as many terms as appear in Ψ, but some are just derivatives of the ones appearing in ϕ and therefore do not represent independent degrees of freedom.

Let us denote $\theta^{\alpha}\bar{\theta}^{\dot{\alpha}}\partial_{\alpha\dot{\alpha}} = U_0$. This operator can be used to implement a change of coordinates in superspace which is often helpful. We note that it generates imaginary translations of the x coordinates by nonlinear terms in $\theta,\bar{\theta}$. If we apply for example e^{-iU_0} to all fields and $e^{-iU_0}\ D\ e^{iU_0}$ to all operators D we obtain a new description where the chiral and antichiral superfields have the form

$$\Phi_c(x,\theta,\bar{\theta}) = \phi(x,\theta) \quad ,$$

$$\bar{\Phi}_c(x,\theta,\bar{\theta}) = e^{-2iU_0}\ \phi^*(x,\bar{\theta}) \quad .$$

At the same time a real supefield $V(x,\theta,\bar{\theta}) = V^*(x,\theta,\bar{\theta})$ goes over to

$$V_c(x,\theta,\bar{\theta}) = e^{-iU_0}\ V(x,\theta,\bar{\theta})$$

and satisfies the new "reality" condition

$$V_C^*(x,\theta,\bar\theta) = e^{2iU_0}\, V_C(x,\theta,\bar\theta) \quad ,$$

while the covariant derivatives take on the form

$$D_{C\alpha} = e^{-iU_0}\, D_\alpha\, e^{iU_0} = i/2(\partial_\alpha + 2i\bar\theta^{\dot\alpha}\, \partial_{\alpha\dot\alpha}) \quad ,$$

$$\bar D_{C\dot\alpha} = e^{-iU_0}\, \bar D_{\dot\alpha}\, e^{iU_0} = -i/2\ \partial_{\dot\alpha} \quad .$$

The advantage of this "chiral representation" is that a chiral field depends only on θ while $\bar D$ is an ordinary derivative, and this facilitates certain calculations. (But any covariant result will be valid in any representation). One can of course also go to an "antichiral representation" by multiplication by e^{+iU_0}.

We have remarked earlier that the various components of a superfield can be obtained trivially by evaluating its θ and $\bar\theta$-derivatives at $\theta = \bar\theta = 0$. An alternative and often much more useful procedure to obtain components is to set $\theta = \bar\theta = 0$ in its covariant D and $\bar D$ derivatives. The resulting components need not be exactly the same, but are related by field redefinitions to the ones obtained by taking ordinary derivatives and therefore represent an equally valid component description of the theory. An obvious advantage of this method (besides being often simpler and "representation" independent) is that differential constraints that the fields satisfy can be directly implemented rather than coming out of explicit detailed algebra. For example for a chiral superfield we define the components by

$$\Phi|_{\theta=0} = a \quad ,$$

$$D_\alpha\Phi|_{\theta=0} = i/2\ \psi_\alpha \quad ,$$

$$D^2\Phi|_{\theta=0} = f \quad ,$$

$$\bar D\Phi|_{\theta=0} = \bar D^2\Phi|_{\theta=0} = \bar D^2 D\Phi|_{\theta=0} = 0 \quad ,$$

$$\bar D_{\dot\alpha}D_\alpha\Phi|_{\theta=0} = \{\bar D_{\dot\alpha}, D_\alpha\}\Phi|_{\theta=0} = i/2\ \partial_{\alpha\dot\alpha}\Phi|_{\theta=0} = i/2\ \partial_{\alpha\dot\alpha}a \quad , \qquad \text{etc.,}$$

where we have used the anticommutation relations and the chirality of Φ. This clearly shows that the number of independent components is small and, while equivalent to expanding $e^{iU_0}[a+\theta\psi+\theta^2 f]$, it is algebraically much simpler and does not require knowledge that a chiral field can be put in this special form.

Invariants and "tensor calculus"

As mentioned earlier supersymmetry transformations are just translations in superspace, and the simplest way to construct invariants is to write down quantities which are translationally invariant. In particular, invariant actions can be constructed as integrals of superfields over superspace. Such actions will be supersymmetric as long as they are translationally invariant, i.e. do not contain explicit x or θ dependence. Conversely the simplest way to write terms which explicitly break supersymmetry is to introduce some explicit θ-dependence in the integrals,[13] . Thus, given superfields $\Phi, V, \ldots$ the following expression is an invariant

$$\int d^4x\, d^4\theta\; P(\Phi, D\Phi, V, D^2V, \ldots) \quad ,$$

where P is an arbitrary (polynomial) function of the superfields and their covariant D and space-time derivatives.

In general the above integral is the only invariant one can construct. An exception is when the integrand is purely chiral or antichiral. In that case the above integral is zero:

$$\int d^4x\, d^4\theta \sim \int d^4x\, d^2\theta\, d\bar{\theta}^{\dot{\alpha}}\, \partial_{\dot{\alpha}} \sim \int d^4x\, d^2\theta\, d\bar{\theta}^{\dot{\alpha}}\, \bar{D}_{\dot{\alpha}} \quad .$$

We have used the important fact that the difference between ∂ and D is a total derivative which we can drop upon x-integration. Then $\bar{D}_{\dot{\alpha}}$ anihilates the chiral integrand it acts upon. (Another way of seeing the same result is to realize that any chiral expression can be written in the form of e^{-iU_0} times an expression depending only on θ. Then, expanding the exponential gives terms which are total derivatives and can be dropped, the integrand has no $\bar{\theta}$ dependence and the integral vanishes). This makes it also clear that a suitable integral to write for chiral integrands is one which involves only a $d^2\theta$ integration. Similarly antichiral expressions are to be integrated with the antichiral measure $d^2\bar{\theta}$. It is easy to check that these are still translationally invariant, i.e. supersymmetric.

We note a variety of ways of representing these integrals (again observing that if one drops total space-time derivatives $\int d\theta_\alpha \sim \partial_\alpha \sim D_\alpha$):

$$\int d^4x\, d^4\theta = \int d^4x\, d^2\theta\, \bar{D}^2 = \int d^4x\, d^2\bar{\theta}\, D^2$$

$$= \int d^4x\, D^2\, \bar{D}^2 = \int d^4x\, D^2\, \bar{D}^2 \big|_{\theta=0} \quad ,$$

where the last step is again a consequence of the fact that the θ-dependent terms are total x-derivatives. This, and the earlier observation that the components of a superfield can be obtained as D derivatives allow perhaps the easiest way of finding supersymmetric component Lagrangians from the corresponding superfield action. For example, the chiral superfield kinetic action

$$I = \int d^4x\, d^4\theta\, \bar{\Phi}\Phi = \int d^4x\, D^2\, \bar{D}^2(\bar{\Phi}\Phi)\big|_{\theta=0}$$

gives

$$L = D^2\left[\left(\bar{D}^2\bar{\Phi}\right)\Phi\right] = \left[\left(D^2\ \bar{D}^2\bar{\Phi}\right)\Phi + 2\left(D^\alpha\ \bar{D}^2\bar{\Phi}\right)\left(D_\alpha\Phi\right) + D^2\Phi\ \bar{D}^2\bar{\Phi}\right]\Big|_{\theta=0} \ .$$

But, on an antichiral superfield, using the anticommutation relations

$$D^\alpha\ \bar{D}^2\bar{\Phi} = i\partial^{\alpha\dot{\alpha}}\ \bar{D}_{\dot{\alpha}}\ \bar{\Phi} \ ,$$

$$D^2\ \bar{D}^2\bar{\Phi} = \Box\ \bar{\Phi} \ ,$$

so that

$$L = -1/2\ \psi_\alpha\ \partial^{\alpha\dot{\alpha}}\ \bar{\psi}_{\dot{\alpha}} + (\Box\ \bar{a})a + \bar{f}f \ ,$$

which, rewritten in terms of A,B,ψ,F,G gives the usual kinetic Lagrangian of the scalar multiplet. This procedure is faster, and less apt to lead to algebraic mistakes than the one which explicitly substitutes the forms of Φ and $\bar{\Phi}$ and evaluates the θ-integrals.

The rest requires very simple algebra. The most general renormalizable action for a chiral superfield (or several) is

$$\int d^4x\ d^4\theta\ \bar{\Phi}\Phi + \int d^4x\ d^2\theta\left(\xi\Phi + 1/2\ m\ \Phi^2 + \lambda/3!\ \Phi^3\right) + \text{h. c.} \ .$$

As another example of the procedure one can imagine working out the interaction term:

$$\int d^4x\ d^2\theta = \int d^4x\ D^2$$

and

$$D^2(\Phi^3)\big|_{\theta=0} = 3\Phi^2 D^2\Phi\big|_{\theta=0} + 6\Phi\ D^\alpha\ \Phi\ D_\alpha\ \Phi\big|_{\theta=0}$$

$$= 3a^2 f - 3/2\ a\psi^2 \ .$$

Other invariant objects one can construct (but do not correspond to renormalizable actions) are made up of

$$\bar{\Phi}^2\Phi^2\ ,\quad \Phi\ D^2\Phi\ ,\quad \Phi^2\ \bar{D}^2\bar{\Phi},\ \text{etc.} \ .$$

Note that the last one is chiral so it would require a $d^2\theta$ integration to produce a non-vanishing result. Quantities involving also a real superfield $V(x,\theta,\bar{\theta})$ (which will be used to describe the vector multiplet) are

$$\bar{\Phi}\ e^V\ \Phi\ ,\quad V\ D^\alpha\ V\ D_\alpha\ V\ ,\quad W^\alpha\ W_\alpha \ ,$$

where the integral with the chiral "field strength" superfield

$$W_\alpha = \bar{D}^2(e^{-V}\ D_\alpha\ e^V) = \bar{D}^2 D_\alpha\ V + \ldots \ ,$$

would again require a chiral integration measure. It is a useful exercise to check that

$$I = \int d^4x\, d^2\theta\, W^\alpha W_\alpha = \int d^4x\, D^2(W^\alpha W_\alpha)|_{\theta=0} \quad ,$$

gives the usual vector multiplet Lagrangian if we define the vector field by

$$A_{\alpha\dot{\alpha}} = [D_\alpha, \overline{D}_{\dot{\alpha}}]\, V(x,\theta,\overline{\theta})|_{\theta=0} \quad ,$$

while

$$\overline{D}^2 D_\alpha\, V|_{\theta=0} = i/2\, \lambda_\alpha \; ,$$

$$D^2 \overline{D}_{\dot{\alpha}}\, V|_{\theta=0} = -i/2\, \overline{\lambda}_{\dot{\alpha}} \; , \quad D^\alpha \overline{D}^2 D_\alpha\, V|_{\theta=0} = D \quad ,$$

define the spinor and auxiliary field of the component theory (in "Wess-Zumino" gauge). Note that the other components of V (which are gauge degrees of freedom) automatically do not contribute to the above action.

As a final item we discuss the "tensor calculus" rules of global supersymmetry: Given chiral superfields Φ, Φ' describing scalar multiplets (α, ψ, f) and (α', ψ', f') the product $\Phi\Phi'$ is again a chiral superfield whose components are to be found from the expansion of the product. This can again be done by writing the superfields out in chiral representation and just doing the multiplication (with $(\theta)^3 = 0$), but it is easier to do it with D derivatives:

$$\Phi\Phi'|_{\theta=0}, \quad D(\Phi\Phi')|_{\theta=0} = (D\Phi)\Phi' + \Phi(D\Phi'), \text{ etc.} \quad ,$$

from which one reads out the "multiplication" rules

$$\left(aa' \; , \; a\psi' + a'\psi \; , \; af' + a'f - 1/2\; \psi'\psi\right) \quad .$$

On the other hand the product $\Phi\overline{\Phi}$ is not chiral, but rather a real (vector multiplet) superfield whose components can again be worked out by applying D's and $\overline{D}$'s to the product.

To make contact with some other component results we observe that the full integration measure appearing in the action just picks out the D-component of the integrand (e.g. of the product $\overline{\Phi}\Phi$) while the chiral measure picks out the F-component of a chiral integrand (e.g. of Φ^3). It is in this context that tensor calculus was used to find invariant actions (especially in supergravity after auxiliary fields were found). All that was needed was to find the right component in the product of components. From a superfield point of view this is a rather unnecessary exercise (provided the superfield formulation is known) since there hardly exists a context in which manipulating the supefields directly is not easier.

Gauge superfields

We shall consider a multiplet of matter fields described by chiral superfields $\Phi = \{\Phi_i\}$ transforming under the action of some global group

$$\Phi \to \Phi' = e^{i\Lambda}\,\Phi \quad ,$$

with constant matrices Λ. Clearly the quantitiy $\bar{\Phi}\Phi = \bar{\Phi}^i\Phi_i$ which appears in the action is invariant if $\bar{\Phi} \to \bar{\Phi}e^{-i\Lambda}$.

We wish to extend Λ to a local quantity. The restriction we shall place on it is that if Φ is chiral so is Φ'. This requires that Λ itself be a chiral superfield. Similarly $\bar{\Phi}' = \bar{\Phi}e^{-i\bar{\Lambda}}$ requires that $\bar{\Lambda}$ be antichiral. But then the kinetic term transform as

$$\bar{\Phi}\Phi \to \bar{\Phi}e^{-i\bar{\Lambda}}\,e^{i\Lambda}\,\Phi$$

and is no longer invariant, requiring the introduction of a gauge field to restore invariance. The simplest guess, which is the correct one, is to introduce a real (hermitean matrix) superfield $V(x,\theta,\bar{\theta})$ with the transformation law

$$e^V \to e^{i\bar{\Lambda}}\,e^V\,e^{-i\Lambda}$$

and observe that $\bar{\Phi}\,e^V\,\Phi$ is then trivially invariant. It is simplest to discuss the Abelian case where $\Lambda,\bar{\Lambda},V$ are single superfields and the transformation law can be rewritten as

$$V \to V + i(\bar{\Lambda}-\Lambda) \quad .$$

We recall the form of the real superfield

$$V = C + \theta\chi + \bar{\theta}\bar{\chi} + \theta^2 M + \bar{\theta}^2 M + \theta\sigma^m\bar{\theta}\,A_m + \ldots \quad .$$

Clearly the chiral superfields $\Lambda,\bar{\Lambda}$ can be used to gauge away part of V (to go to "Wess-Zumino" gauge) by essentially cancelling the first few components. The remainder contains just the usual component fields of the vector multiplet: A_m,λ_α,D. Effectively we have

$$V = \theta\sigma^m\bar{\theta}\,A_m + \bar{\theta}^2\theta\lambda + \theta^2\bar{\theta}\bar{\lambda} + \theta^2\bar{\theta}^2 D$$

and it is easy to verify then that the expansion into components of the covariantized kinetic term $\bar{\Phi}\,e^V\,\Phi$ gives the usual gauge-covariantized interaction between the scalar and (Abelian) vector multiplet. Again the best way is not by brute force expansion but by using D derivatives and the Wess-Zumino gauge

$$V|_{\theta=0} = DV|_{\theta=0} = \bar{D}V|_{\theta=0} = D^2V|_{\theta=0} = \bar{D}^2V|_{\theta=0} = 0 \quad .$$

It is also clear that $W_\alpha = \bar{D}^2 e^{-V} D_\alpha e^V$ is a gauge invariant object (use the chiralities of $\Lambda, \bar{\Lambda}$ to pull them through the derivatives) and can be used to construct the vector multiplet action as mentioned in the previous section.

The non-Abelian case is not different except for noncommutativity of the various objects. For example one must define $\Lambda = \Lambda_i \tau^i$ where τ^i represent the generators of the gauge group in the representation of the Φ's while $V = V_i T^i$ where the T's are the generators in the adjoint representation. The gauge transformations are highly nonlinear and one defines now

$$W_\alpha = [\bar{D}^{\dot{\alpha}}, \{\bar{D}_{\dot{\alpha}} , e^{-V} D_\alpha e^V\}] .$$

It is of course desirable to define also (gauge) covariant derivatives which we denote by $\nabla_\alpha, \bar{\nabla}_{\dot{\alpha}}$, and $\nabla_{\alpha\dot{\alpha}}$. We begin by requiring that $\Phi \to e^{i\Lambda}\Phi$ should imply $\nabla_A \Phi \to e^{i\Lambda} \nabla_A \Phi$ and it is easy to see that

$$\nabla_\alpha = e^{-V} D_\alpha e^V ,$$

$$\bar{\nabla}_{\dot{\alpha}} = \bar{D}_{\dot{\alpha}} ,$$

$$i/2 \ \nabla_{\alpha\dot{\alpha}} = \{\nabla_\alpha, \bar{\nabla}_{\dot{\alpha}}\} ,$$

will work when acting on a chiral superfield. Similarly when acting (from the right) on an antichiral superfield, suitable covariant derivatives are (cf. $D_m = \partial_m \pm ieA_m$ where the sign depends on the charge of the field on which D_m is acting)

$$\overleftarrow{\nabla}_\alpha = \overleftarrow{D}_\alpha ,$$

$$\overleftarrow{\bar{\nabla}}_{\alpha} = e^V \overleftarrow{\bar{D}}_{\dot{\alpha}} e^{-V} ,$$

$$i/2 \ \overleftarrow{\nabla}_{\alpha\dot{\alpha}} = \{\overleftarrow{\nabla}_a, \overleftarrow{\bar{\nabla}}_{\dot{\alpha}}\} .$$

There exists a more geometrical approach to this subject which seems a useful introduction to the corresponding situation in supergravity. In general, if one has postulated the existence of a set of covariant derivatives

$$\nabla_A = \{\nabla_\alpha , \bar{\nabla}_{\dot{\alpha}} , \nabla_{\alpha\dot{\alpha}}\} ,$$

associated with some local group of transformations one can define torsions and field strenghts (curvatures) in terms of the (anti)commutators

$$[\nabla_A, \nabla_B\} = T_{AB}{}^C \nabla_C + F_{AB} .$$

One may expect to find the usual field strenghts among the components of F_{AB}.

The torsions and curvatures satisfy a certain number of constraints (Bianchi identities) which follow from the Jacobi identities $\sum [\nabla_A, [\nabla_B, \nabla_C\}\} = 0$. If one writes

$$\nabla_A = D_A - i\Gamma_A \quad ,$$

with connection terms Γ_A, one may expect to be able to express them in terms of suitably defined (unconstrained) gauge fields. In general these gauge fields will not describe irreducible representations of supersymmetry, and one way to obtain irreducibility is to impose further constraints on the torsions and curvatures beyond those that follow from the Bianchi identities. In general imposing constraints and solving them is an art but in this case (and in some cases for supergravity) the artists have done their work and we know how to proceed.

It turns out that the correct constraints to impose in this case is to require that all components of the torsion be zero except

$$T_{\alpha\dot{\alpha}}{}^{a} = i/2\ \sigma^{a}{}_{\alpha\dot{\alpha}} \quad .$$

In addition one requires

$$F_{\alpha\dot{\alpha}} = F_{\alpha\beta} = F_{\dot{\alpha}\dot{\beta}} = 0 \quad .$$

What are the implications of these relations? First, the choice of $T_{\alpha\dot{\alpha}}{}^{a}$ and $F_{\alpha\dot{\alpha}}$ means that

$$\{\nabla_\alpha, \overline{\nabla}_{\dot{\alpha}}\} = i/2\ \nabla_{\alpha\dot{\alpha}} \quad ,$$

which just generalizes the flat space construction of the space-time derivative from the anticommutator of spinor derivatives. Second, the choice $F_{\alpha\beta} = F_{\dot{\alpha}\dot{\beta}} = 0$ means for example that we can define "covariantly chiral" and antichiral superfields, e.g.

$$\overline{\nabla}_{\dot{\alpha}}\tilde{\Phi} = 0, \qquad \nabla_\alpha\overline{\tilde{\Phi}} = 0 \quad ,$$

since this is required by

$$0 = \{\overline{\nabla}_{\dot{\beta}}, \overline{\nabla}_{\dot{\alpha}}\}\tilde{\Phi} = F_{\dot{\beta}\dot{\alpha}}\tilde{\Phi} \quad .$$

One now proceeds in two directions: Either one examines the Bianchi identities to show that they can still be satisfied (the constraints above are not so severe as to imply only a trivial solution) and "solves" them by showing that the remaining components of F_{AB} can all be expressed in terms of a single chiral superfield W_α (the field strength superfield), thus showing that the constraints have given us an irreducible representation, <u>or</u> one really solves the constraints, i.e. shows that one can write the covariant derivatives in terms of unconstrained gauge fields such that the constraints are automatically satisfied. To indicate how this goes let us work out what the constraints imply for some of the connections. We consider for simplicity the Abelian case. We have, with $\nabla_A = D_A - i\Gamma_A$, since $T_{\alpha\beta}{}^C = F_{\alpha\beta} = 0$

$$i\{\nabla_\alpha, \nabla_\beta\} = D_\alpha \Gamma_\beta + D_\beta \Gamma_\alpha = 0 \quad .$$

This relation implies that for some general superfield W (not to be confused with W_α)

$$\Gamma_\alpha = -iD_\alpha W \quad .$$

Similarly

$$\Gamma_{\dot\alpha} = i\overline{D}_{\dot\alpha} \overline{W}$$

and it is then obvious that one can write

$$\nabla_\alpha = e^{-W} D_\alpha e^W \quad ,$$

$$\overline{\nabla}_{\dot\alpha} = e^{\overline{W}} \overline{D}_{\dot\alpha} e^{-\overline{W}} \quad ,$$

with all the other constraints satisfied. Thus the covariant derivatives and all other objects can be expressed in terms of the general superfield W.

For completeness we should make contact with the previous formulation. (This is not totally elementary and could be skipped). We have introduced the notion of a covariantly chiral superfield $\overline{\nabla}_{\dot\alpha}\tilde\Phi = 0$. Obviously we can express it as

$$\tilde\Phi = e^{\overline{W}} \Phi \quad ,$$

where Φ is ordinary chiral. But now

$$\nabla_\alpha \tilde\Phi = e^{-W} D_\alpha e^W e^{\overline{W}} \Phi$$

and if this object is to transform covariantly we may demand that at least

$$e^W \to e^W e^{-iK}, \qquad e^{-W} \to e^{iK} e^{-W}, \qquad e^{\overline{W}} \to e^{iK} e^{\overline{W}} \quad ,$$

where K is a real superfield. On the other hand we still want Φ to transform as

$$\Phi \to e^{i\Lambda} \Phi, \text{ etc.} \quad .$$

The upshot is that we must assume transformation laws

$$e^W \to e^{i\overline{\Lambda}} e^W e^{iK}, \qquad e^{\overline{W}} \to e^{-iK} e^{\overline{W}} e^{-i\Lambda} \quad ,$$

$$\Phi \to e^{i\Lambda}\,\Phi, \qquad \bar{\Phi} \to \bar{\Phi} e^{-i\bar{\Lambda}} \quad ,$$

and therefore

$$\tilde{\Phi} \to e^{-iK}\,\tilde{\Phi}, \quad \bar{\tilde{\Phi}} \to \bar{\tilde{\Phi}}\, e^{iK} \quad ,$$

while the gauge covariant derivatives transform simply as

$$\nabla_A \to e^{-iK}\,\nabla_A\, e^{iK} \quad ,$$

independently of Λ.

We observe that this is a larger gauge group than previously discussed. However one can use the K-transformations to gauge away the imaginary part of W and $\bar{W}$. In that gauge (which is maintained by making further K-transformations for each Λ-transformation) $W = \bar{W} = V/2$ and we recover the previous results with the restricted gauge group of transformations parametrized by Λ, $\bar{\Lambda}$. The situation in supergravity is very similar.

To conclude this section we note that in addition to θ-expansion and applying D-derivatives, there is another way in which matter (chiral superfield)-Yang-Mills (gauge superfield) component couplings can be obtained. In terms of covariantly chiral fields, the Lagrangian $\bar{\Phi}\, e^V\, \Phi$ can be written as $\bar{\tilde{\Phi}}\tilde{\Phi}$ and the θ-integration can be replaced by

$$\nabla^2 \bar{\nabla}^2 (\bar{\tilde{\Phi}}\tilde{\Phi})\big|_{\theta=0} \quad .$$

It turns out that up to field redefinitions we can now define the components by $a = \tilde{\Phi}|_{\theta=0}$, $i/2\ \psi_\alpha = \nabla_\alpha \tilde{\Phi}|_{\theta=0}$, etc. The new feature is that in evaluating the expression above, the commutators of the covariant derivatives will contain now the usual covariant space-time derivative $\nabla_{\alpha\dot{\alpha}}$ and the covariant superfield strength W_α, with

$$i/2\ \lambda_\alpha = W_\alpha\big|_{\theta=0} \quad \text{and} \quad D = \nabla^\alpha W_\alpha\big|_{\theta=0} \quad .$$

In this way the coupling to the vector multiplet is made explicit.

N=1 Supergravity

At the present time we do not have a complete superfield description of all (N=1,2,...8) supergravity theories. This is equivalent to saying in component language that we do not know all the auxiliary fields. On the other hand the situation is quite clear for simple supergravity which we now discuss.

Supergravity was originally defined in terms of the graviton and the gravitino fields $h_{mn} = g_{mn} - \delta_{mn}$ (or $1/2\ h_{\alpha m} = e_{\alpha m} - \delta_{\alpha m}$) and $\psi_{m\alpha}$. It was recognized early that auxiliary fields had to be adjoined to these in order to obtain a closed algebra of local supersymmetry transformations. One expects that in a superfield formulation some additional gauge degrees of freedom might have to be introduced (cf. the vector multiplet superfield). In fact there is a great deal of gauge freedom, which can be confusing, but can also be put to good use in simplifying calculations.

The basic object of supergravity is a real axial-vector superfield H^a which has the same expansion as V, but with each component bearing an extra vector index a:

$$H^a(x,\theta,\overline{\theta}) = \ldots + \theta\sigma^b\overline{\theta}\, h_b{}^a + \overline{\theta}^2\theta\psi^a + \theta^2\overline{\theta}\overline{\psi}^a + \theta^2\overline{\theta}^2 A^a \quad .$$

In addition one may introduce a chiral superfield (in chiral representation)

$$\chi = \ldots + \theta^2(S + iP) \quad ,$$

which plays a role similar to that of the scalar $\phi(x)$ one introduces in Einstein gravity to make it formally scale invariant. In a suitable (Wess-Zumino type) gauge, all but the components shown can be set to zero. A^a, S, and P are the auxiliary fields of minimal supergravity.

There are several ways to motivate the introduction of these superfields and the subsequent developments, which are similar to those for the case of Yang-Mills systems. One way is o try to write actions for matter (chiral and gauge) superfields which are invariant under local coordinate transformations (in superspace). Another is to construct covariant derivatives, with torsions and curvatures satisfying suitable constraints, and to solve these constraints in terms of unconstrained gauge fields. In both approaches one also has to learn how to write an action for the supergravity superfields themselves, and how to (if so desired) efficiently obtain the corresponding component field results (invariant Lagrangians, local "tensor calculus", etc.). I will not attempt to derive the results that follow and simply hope that the analogous results for the Yang-Mills gauge multiplet make them not too unreasonable.

In analogy to the Yang-Mills situation we start with chiral superfields and we introduce local translations (general coordinate transformations) which preserve the chirality:

$$\phi \to e^{\Lambda}\,\phi, \quad \partial_{\dot{\nu}}(e^{\Lambda}\,\phi) = 0 \quad ,$$

$$\Lambda(x,\theta,\overline{\theta}) = \Lambda^m\,\partial_m + \Lambda^\mu\,\partial_\mu + \Lambda^{\dot{\mu}}\,\partial_{\dot{\mu}} \quad .$$

Therefore the chirality preserving condition is

$$\partial_{\dot{\nu}} \Lambda^m = \partial_{\dot{\nu}} \Lambda^\mu = 0 \quad .$$

Note that we use ordinary derivatives $\partial_\nu, \partial_{\dot{\nu}}$, so that chirality means simply independence of $\bar{\theta}$. These are not covariant objects but neither would be $D_\alpha, \bar{D}_{\dot{\alpha}}$, and one might as well start with something simpler. We define antichiral fields in a similar fashion, transforming with objects $\bar{\Lambda}$ which satisfy the restrictions

$$\partial_\nu \bar{\Lambda}^m = \partial_\nu \bar{\Lambda}^{\dot{\mu}} = 0 \quad .$$

Note that the superfields Λ^μ and $\bar{\Lambda}^\mu$ satisfy no constraints.

The analogy with Yang-Mills would be to recognize that $\phi\bar{\phi}$ is not invariant and to introduce a gauge field (operator)

$$U = U^m \partial_m + U^\mu \partial_\mu + U^{\dot{\mu}} \partial_{\dot{\mu}} \quad ,$$

transforming as

$$e^{-2iU} \rightarrow e^{\Lambda} e^{-2iU} e^{-\bar{\Lambda}} \quad ,$$

to try to make $\phi\, e^{-2iU} \bar{\phi}$ invariant. However, because Λ is a differential operator this is insufficient.

The local gauge group defined by Λ is huge and certainly contains more than what we want (just count the number of free parameters). One can restrict it as follows: With some algebra (details of what follows can be found in Ref. 9) one can show that $\Lambda^{\dot{\mu}}, \bar{\Lambda}^\mu$ can be used to gauge away U^μ and $U^{\dot{\mu}}$. Then one insists that further transformations should keep us in this gauge. This is a restriction on Λ which eliminates almost all of the unwanted part of the group. What is left are first of all some transformations which could be used (if so desired) to take the remaining U^m to a Wess-Zumino type gauge in which the first few components are zero. In general, care must be taken to define the remaining components so that they transform covariantly under the residual symmetries. These consist of local space-time and supersymmetry transformations (in the sense of component supergravity) and superscale (conformal) transformations. These last could be eliminated, but it turns out to be more fruitful to keep them and introduce a chiral compensating field χ transforming suitably under these scale transformations, and retain the freedom associated with the ensuing scale invariance. The scale can be fixed at any stage by giving χ a fixed value.

The chiral field χ can be defined in such a way that the chiral measure

$$\int d^4x \, d^2\theta \, \chi^3$$

is invariant (χ is a density) and can be used to write invariant mass or chiral interaction terms

$$\int d^4x\, d^2\theta\, \chi^3 [1/2\, m\phi^2 + \lambda/3!\, \phi^3] \quad .$$

On the other hand the full integral $\int d^4x\, d^4\theta\, \phi\, e^{-2iU}\, \overline{\phi}$ is not invariant. What is missing is the analog of e = g. We need to do a bit more geometry, introduce a superspace "achtbein" (the analog of the usual vierbein) and a superdeterminant E, functions of U and χ, and show that with this determinant

$$\int d^4x\, d^4\theta\, E\, \phi\, e^{-2iU}\, \overline{\phi}$$

is invariant. It will turn out that the same E is the proper object to describe the gauge action itself.

To go on one introduces the notion of covariant derivative. The standard procedure is to introduce local frames and define the achtbein $E_M{}^A$, relating the world coordinates indexed by $M = (m,\mu,\dot{\mu})$ to the tangent space coordinates indexed by $A = (a,\alpha,\dot{\alpha})$. One also defines local Lorentz transformations and restricts the supergravity group by requiring that the vector and spinor coordinates transform in standard fashion, without mixing, by means of the same Lorentz transformation. Therefore the action of the Lorentz group on the eight dimensional tangent superspace is reducible.

The (inverse) achtbein can be used to define a tangent vector field

$$E_A = E_A{}^m\, \partial_m + E_A{}^\mu\, \partial_\mu + E_A{}^{\dot{\mu}}\, \partial_{\dot{\mu}} \quad .$$

Note that at this stage one could just as well, by a rearrangement of terms express this in terms of D-derivatives

$$E_A = E_A{}^B\, D_B = E_A{}^b\, \partial_b + E_A{}^\beta\, D_\beta + E_A{}^{\dot{\beta}}\, \overline{D}_{\dot{\beta}} \quad ,$$

which is very convenient for exhibiting the global supersymmetry of the theory.

The covariant derivatives are defined now in standard fashion:

$$\nabla_A = E_A + \Phi_A = E_A{}^B\, D_B + \Phi_A \quad ,$$

where the (Lorentz) connection terms Φ_A are introduced as usual to cancel derivatives of gauge parameters and thus insure covariance. Their action on vector and spinor fields is

$$(\Phi_A\, v)^b = \Phi_A{}^b{}_c\, v^c$$

$$(\Phi_A\, \lambda)^\beta = -\Phi_A{}^\beta{}_\gamma\, \lambda^\gamma$$

where $\Phi_{Aa}{}^{b}$ $(A = a, \alpha, \dot{\alpha})$ are connection coefficients, functions of the basic fields, and

$$\Phi_{A\beta}{}^{\gamma} = 1/2\ \sigma^{a}{}_{\beta\dot{\gamma}}\ \sigma_{b}{}^{\gamma\dot{\gamma}}\ \Phi_{Aa}{}^{b} \quad .$$

The fact that only $\Phi_{Aa}{}^{b}$ are independent while $\Phi_{A\alpha}{}^{\beta}$ (and similarly $\Phi_{A\dot{\alpha}}{}^{\dot{\beta}}$) are defined in terms of them is related to the earlier remark that the action of the Lorentz group is reducible.

Torsions and curvatures are defined by

$$[\nabla_{A}, \nabla_{B}\} = T_{AB}{}^{C}\ \nabla_{C} + 1/2\ R_{AB} \quad ,$$

where R_{AB} has components $R_{ABc}{}^{d}$, $R_{AB\alpha}{}^{\beta}$, and $R_{AB\dot{\alpha}}{}^{\dot{\beta}}$ (when acting on vector or spinor indices), these last two being related to the first one in the same way that the components of Φ_{A} are related. Note that from now on everything is referred to tangent space coordinates. Since this can be expressed in terms of flat superspace, we are on familiar grounds as far as algebraic manipulations are concerned.

The torsions and curvatures are restricted by Bianchi identities. They have also been restricted already by the way the spinor components of connections and curvatures are related to the vector components. Further restrictions are necessary to insure irreducibility of the supersymmetry representation. These are known as Wess-Zumino constraints and are given by

$$T_{\alpha\dot{\beta}}{}^{c} = 1/2\ i\sigma^{c}{}_{\alpha\dot{\beta}} \quad ,$$

$$T_{\alpha\beta}{}^{\gamma} = T_{\alpha\beta}{}^{\dot{\gamma}} = T_{\alpha\beta}{}^{c} = T_{ab}{}^{c} = R_{\alpha\dot{\beta}c}{}^{d} = 0 \quad ,$$

with the other components unrestricted (This is a slight modification of the original Wess-Zumino constraints. It is adapted to the Siegel-Gates formalism but has the same effect as the original constraints).

Just as in Yang-Mills theory one can proceed in two directions. From the Bianchi identities one can deduce that the remaining components of curvature and torsions can all be expressed in terms of a few basic superfields. The analogue of the superfield strength $W_{\alpha} = \overline{D}^{2}e^{-V}\ D_{\alpha}\ e^{V}$ consists now of three superfields denoted by R, $G_{\alpha\dot{\beta}}$, $W_{\alpha\beta\gamma}$ (the latter is chiral and totally symmetric in its indices). The result is best presented by writing now the commutators of covariant derivatives in terms of them:

$$\{\nabla_{\alpha}, \nabla_{\beta}\} = -\overline{R}\ M_{\alpha\beta} \quad ,$$

$$\{\nabla_{\alpha}, \overline{\nabla}_{\dot{\beta}}\} = i/2\ \nabla_{\alpha\dot{\beta}} \quad ,$$

$$[\nabla_\alpha, i/2\ \nabla_{\beta\dot\gamma}] = -1/2\ \varepsilon_{\alpha\beta}[-\overline{R}\ \overline{\nabla}_{\dot\gamma} + G^\delta{}_{\dot\gamma}\ \nabla_\delta + (\nabla^\delta\ G^\varepsilon{}_{\dot\gamma})M_{\delta\varepsilon} + \overline{W}_{\dot\gamma\dot\delta}{}^{\dot\varepsilon}\ \overline{M}_{\dot\varepsilon}{}^{\dot\delta}] -1/2(\overline{\nabla}_{\dot\gamma}\ \overline{R})M_{\alpha\beta} \quad ,$$

where the bar denotes complex conjugation and the M's are the infinitesimal generators of the Lorentz group

$$M_{\alpha\beta}\ \lambda^\gamma = -1/2(\delta^\gamma{}_\alpha\ \lambda_\beta + \delta^\gamma{}_\beta\ \lambda_\alpha) \quad .$$

Of the above equations the second one generalizes the flat space relation, the right hand side being the usual spacetime covariant derivative. The others are useful both as defining equations for the superfield strengths, and for finding component forms for the supergravity-matter couplings. Additional useful relations are

$$\overline{\nabla}^{\dot\beta}\ G_{\alpha\dot\beta} = \nabla_\alpha\ R\ , \qquad \nabla^\alpha\ W_{\alpha\beta\gamma} = i/4\ \nabla_{(\beta\dot\alpha}\ G_{\gamma)}{}^{\dot\alpha} \quad .$$

The superfields R, $G_{\alpha\dot\beta}$, $W_{\alpha\beta\gamma}$ can be written in terms of the basic gauge potentials of the theory. Thus ultimately they can be expressed in terms of the graviton and gravitino fields. In addition the supergravity field equations take on a very simple form

$$R = G_{\alpha\dot\beta} = \nabla^\alpha\ W_{\alpha\beta\gamma} = 0 \quad .$$

Instead of trying to satisfy the Bianchi identities, one can proceed in the other direction, actually finding the unconstrained (gauge) superfields (they will turn out to be H^a and χ) in terms of which the covariant derivatives (i.e. the quantities $E_A{}^B$ and $\Phi_{AB}{}^C$) can be expressed, so that the Wess-Zumino constraints are automatically satisfied. The solution of these constraints was found by W. Siegel,[14] and is rather involved. To try and follow the details in Ref. [8] may be a bit rough. You might consult Ref. [9] and the appendix of "Supergravity I".[15]

The solution to the constraints involves first expressing the superfield U in terms of the axial supervector superfield H^A by splitting off a flat superspace piece:

$$e^{-2iU} = e^{-i\theta^\alpha\overline{\theta}^{\dot\alpha}\ \partial_{\alpha\dot\alpha}}\ e^{-iH^A\ D_A}\ e^{-i\theta^\alpha\overline{\theta}^{\dot\alpha}\ \partial_{\alpha\dot\alpha}} \quad ,$$

which is analogous to the splitting $e_{am} = \delta_{am} + 1/2\ h_{am}$. Note that in this way (recall we defined chiral superfields ϕ by taking them independent of $\overline{\theta}$)

$$\phi(x,\theta)\ e^{-2iU}\ \overline{\phi}(x,\overline{\theta}) = \Phi(x,\theta,\overline{\theta})\ e^{-iH}\ \overline{\Phi}(x,\theta,\overline{\theta}) \quad ,$$

where the $\Phi,\overline{\Phi}$ are now chiral and antichiral supefields in <u>vector</u> representations with respect to the global supersymmetry algebra of the D's in flat superspace. Recall that the Λ-group could be used to gauge away $U^\mu, \dot U^\mu$; it can be used just as well to gauge away

$H^{\alpha}, H^{\dot{\alpha}}$, leaving us with

$$H = H^a \partial_a \quad ,$$

where H^a is the axial vector superfield introduced at the beginning of this section.

The covariant derivatives $\nabla_A = E_A + \Phi_A$ can now be expressed in terms of H and χ: In a particular Lorentz frame when acting on a chiral superfield (cf. the Yang-Mills case)

$$E_{\alpha} = \Psi(H,\chi)\; e^{-iH}\, D_{\alpha}\, e^{iH} \quad ,$$

$$E_{\dot{\alpha}} = \overline{\Psi}(H,\chi)\; \overline{D}_{\dot{\alpha}} \quad .$$

However, Ψ, $E_{\alpha\dot{\alpha}}$, and Φ_A as well as

$$E = \text{superdet}\left[E^A{}_M\right] \quad ,$$

are somewhat complicated functions of H and χ and will not be given here. But it should be emphasized that everything is explicit, everything is written in terms of flat superspace superfields H^a, χ and derivatives $D_{\alpha}, \overline{D}_{\dot{\alpha}}, \partial_a$ and, except for the fact that all expresions are much more complicated than in the global supersymmetry case, the algebra is exactly the same. The kinetic action for a chiral superfield is

$$\int d^4x\, d^4\theta\; E\, \Phi\, e^{-iH}\, \overline{\Phi} \quad ,$$

(we have given earlier the mass and cubic coupling terms) while the action for a vector multiplet is

$$\int d^4x\, d^2\theta\; (\chi)^3\; W^{\alpha}\, W_{\alpha} \quad ,$$

where

$$W_{\alpha} = (\overline{\nabla}^2 + R)\left[e^{-V}\, \nabla_{\alpha}\, e^{V}\right] \quad ,$$

is the properly covariantized field strength for the real gauge superfield V. Note the presence of the curvature superfield $R(H,\chi)$ which was introduced earlier in the solution of the Bianchi identities. Its role is to insure that W_{α} is chiral, since one can show from the commutation relations that when acting on a superfield with <u>no</u> dotted indices, $\overline{\nabla}(\overline{\nabla}^2 + R) = 0$ gives the generalization of the flat superspace relation $(\overline{D})^3 = 0$. Finally the action for supergravity itself is simply

$$\int d^4x\, d^4\theta\; E(H,\chi) \quad .$$

The above actions can be expanded in powers of H and are suitable for superfield perturbation theory calculations,[15 10]. However, for expansion into components it is often more convenient to replace the θ-integration by D-derivatives or, even better, by covariant derivatives ∇. After a certain amount of gymnastics one can show that the following are alternative expressions for quantities appearing in the action:

$$\int d^4x\, d^2\theta\, \chi^3 = \int d^4x\, d^4\theta\, E/R \quad ,$$

$$\int d^4x\, d^4\theta\, E = \int d^4x\, d^2\theta\, \chi^3\, (\overline{\nabla}^2 + R) \quad ,$$

$$= \int d^4x\, e\left[\nabla^2 + \overline{\psi}_m{}^{\dot\alpha}\, \sigma^{m\alpha}{}_{\dot\alpha}\, \nabla_\alpha + 3(S + iP) - \overline{\psi}_m\, \sigma^{mn}\, \psi_n\right](\overline{\nabla}^2 + R) \quad ,$$

where

$$e = \det\left[E_m{}^a\right]\Big|_{\theta=0} \quad ,$$

$$\overline{\psi}_m{}^{\dot\alpha} = 2E_m{}^{\dot\alpha}\Big|_{\theta=0} \quad .$$

One can proceed now just like in flat superspace, acting with the covariant derivatives on the superfields and defining the components as values of the derivatives at $\theta = 0$,[5 16]. Thus, for example, we write for a covariantly chiral superfield $\tilde\Phi$ ($\overline{\nabla}\tilde\Phi = 0$)

$$a = \tilde\Phi\Big|_{\theta=0}, \quad i/2\ \psi_\alpha = \nabla_\alpha \tilde\Phi\Big|_{\theta=0}, \quad f = \nabla^2\tilde\Phi\Big|_{\theta=0} \quad ,$$

where we have simply defined $\tilde\Phi$ by $\int E\, \Phi\, e^{-iH}\, \overline{\Phi} = \int E\, \tilde\Phi\tilde{\overline{\Phi}}$. Similary for the vector multiplet we define components in analogy to the global supersymmetry case.

In the course of this procedure we will encounter commutators of ∇'s which will introduce the superfields R, $G_{\alpha\dot\beta}$, $W_{\alpha\beta\gamma}$. All we need to know is that in a particular gauge the achtbein itself has components at $\theta = 0$

$$E_m{}^a = e_m{}^a, \qquad E_m{}^\alpha = 1/2\ \psi_m{}^\alpha \quad ,$$

$$E_\mu{}^a = 0, \qquad E_\mu{}^\alpha = \delta^\alpha{}_\mu \quad ,$$

$$\partial^\mu\, E_\mu{}^B = 0 \quad ,$$

where $e_m{}^a, \psi_m{}^\alpha$ are the usual vierbein and gravitino fields, and

$$\Phi_{\mu a}{}^b = 0 \quad ,$$

while $\Phi_{m\alpha}{}^b$ is the space-time connection. Furthermore,

$$G_{\alpha\dot\beta}\Big|_{\theta=0} = A_{\alpha\dot\beta} \quad , \qquad R\Big|_{\theta=0} = S + iP \quad ,$$

give the auxiliary fields. Finally the chiral $W_{\alpha\beta\gamma}$ and its derivatives contain the gravitino field strenght and the Weyl tensor:

$$W_{\alpha\beta\gamma}|_{\theta=0} = \psi_{\alpha\beta\gamma} , \qquad \nabla_{(\alpha} W_{\beta\gamma\delta)} = C_{\alpha\beta\gamma\delta} , \text{ etc.} \quad .$$

Following this procedure the component couplings of the scalar multiplet and the vector multiplet to supergravity, as well as the component supergravity action itself can be obtained. However, there seems to be little purpose in doing this when calculations with superfields are so much easier. For example the calculation of the (finite) one loop four-particle scattering amplitude in N = 8 supergravity is basically a five minute affair.[17]

Most of my knowledge of the subject comes from discussions with Jim Gates and Warren Siegel and I would like to thank them for putting up with my slow pace. I also would like to thank Martin Roček for sharing with me his understanding of the subject, Martin Roček for keeping me honest, and especially Martin Roček for help with the typing of these lectures.

References

1. A. Salam and J. Strathdee, Nucl. Phys. B76, 477 (1974)
2. J. Wess and B. Zumino, Nucl. Phys. B70, 39 (1974)
3. A. Salam and J. Strathdee, Fort. der Phys. 26, 57 (1978)
4. P. Fayet and S. Ferrara, Phys. Reports 32 C, 1 (1977)
5. B. Zumino, in "Recent Developments in Gravitation, Cargese 1978", eds. M. Levy and S. Deser, Plenum Press, N.Y.
6. J. Wess, in "Topics in Quantum Field Theory and Gauge Theories, Salamanca 1977", Phys. 77, Springer Verlag, Berlin
7. R. Grimm, in "An Introduction to Supergravity" (Trieste 1981), eds. S. Ferrara and J.G. Taylor, Cambridge University Press (to be published)
8. W. Siegel and S.J. Gates, Nucl. Phys B147, 77 (1979)
9. M. Roček, in "Superspace and Supergravity" (Cambridge 1980), eds. S.W. Hawking and M. Roček, Cambridge University Press, Cambridge
10. M.T. Grisaru, in "An Introduction to Supergravity" (Trieste 1981), eds. S. Ferrara and J.G. Taylor, Cambridge University Press (to be published)
11. E. Sokatchev, Nucl. Phys. B99, 96 (1975)
12. W. Siegel and S.J. Gates, Nucl. Phys. B189, 295 (1981)
13. L. Girardello and M.T. Grisaru, Nucl. Phys. (to be published)
14. W. Siegel, Nucl. Phys. B142, 301 (1978)
15. M.T. Grisaru and W. Siegel, Nucl. Phys. B187, 149 (1981)
16. J. Wess and B. Zumino, Phys. Lett. 79 B, 394 (1978)
17. M.T. Grisaru and W. Siegel (to be published)

GRAND UNIFICATION WITH AND WITHOUT SUPERSYMMETRY

Edward Witten*

Joseph Henry Laboratories
Princeton University
Princeton, New Jersey 08544

These lectures notes provide a general introduction to grand unified theories and their supersymmetric generalizations.

I. Introduction

These lectures will mainly deal with a few phenomenological aspects of supersymmetric grand unified theories. But perhaps it would be useful to first review some properties of ordinary grand unified theories. I will concentrate, for simplicity, on the Georgi-Glashow SU(5) model[1]. I will not attempt a systematic review, but only mention those properties that we will need.

Prior to 1963, there was in particle physics the SU(2) symmetry of isospin conservation, originally discovered in studies of pions and nucleons. There were also Abelian conservation laws, conservation of hypercharge and of baryon number. Gell-Mann and Ne'eman then suggested that the SU(2) x U(1) group of isospin and hypercharge conservation could be unified into a larger approximate symmetry group, SU(3), of the strong interactions. The embedding of SU(2) x U(1) in SU(3) is simplicity itself. A generator of SU(3) is a traceless, hermitian 3x3 matrix. The isopin matrices I are traceless, hermitian 2x2 matrices which can be embedded in a 3x3 matrix by placing them in the upper left-hand corner

* Supported in part by NSF Grant PHY80-19754.

$$\left[\begin{array}{c|c} \vec{I} & 0 \\ \hline 0 & 0 \end{array}\right] \quad . \tag{1}$$

The hypercharge operator Y must commute with I, and this (together with the fact that, like any generator of SU(3), Y must be traceless) determines Y uniquely, apart from normalization

$$Y = \left[\begin{array}{cc|c} 1/3 & & \\ & 1/3 & \\ \hline & & -2/3 \end{array}\right] \quad . \tag{2}$$

The normalization is conventional.

As we now understand it, the SU(3) symmetry that I have just described, which is the symmetry of the "eightfold way" (to be distinguished from the color SU(3) symmetry, discovered later) is less fundamental than the objects on which it acts. Those objects are the light quarks, u, d, and s. Arranging them as a three component column vector

$$\left[\begin{array}{c} u \\ d \\ \hline s \end{array}\right] \quad , \tag{3}$$

which is acted on by the 3x3 SU(3) matrices, we see that the isospin group rotates the u an d quarks among themselves, while the full SU(3) symmetry relates all three light quarks.

The Georgi-Glashow SU(5) model deals in a rather analogous way with the problem of unifying the gauge symmetries which are the basis for our current understanding of particle physics. According to QCD, the strong interactions are described by a local SU(3) gauge symmetry. The Weinberg-Salam model describes the weak and electromagnetic interactions by a local gauge symmetry SU(2) x U(1). The SU(5) model, which is based on the simple fact that 3+2 = 5, embeds SU(3) x SU(2) x U(1) in SU(5) in the following way. An SU(5) generator is a traceless, hermitian 5x5 matrix. The generators λ^a (a = 1...8) of SU(3) are placed in the upper left-hand 3x3 block

$$\left[\begin{array}{c|c} \lambda^a & 0 \\ \hline 0 & 0 \end{array}\right] \quad . \tag{4}$$

The SU(2) matrices $\vec{T}$ are placed in the lower right 2x2 block

$$\left[\begin{array}{c|c} 0 & 0 \\ \hline 0 & \vec{T} \end{array}\right] \quad . \tag{5}$$

The weak hypercharge operator Y is (apart from normalization) the unique traceless matrix that commutes with (4) and (5),

$$Y = \left[\begin{array}{ccc|cc} 2/3 & & & & \\ & 2/3 & & & \\ & & 2/3 & & \\ \hline & & & -1 & \\ & & & & -1 \end{array}\right] \quad . \tag{6}$$

Of course, embedding SU(3) x SU(2) x U(1) in SU(5) is by no means enough to ensure a phenomenologically acceptable model. We also must, among other things, assign the quarks and leptons to SU(5) representations. The quarks and leptons, as we now know them, appear to form three "families"

$$\begin{bmatrix} u \\ d \\ e \\ \nu_e \end{bmatrix} , \quad \begin{bmatrix} c \\ s \\ \mu \\ \nu_\mu \end{bmatrix} \quad \text{and} \quad \begin{bmatrix} t \\ b \\ \tau \\ \nu_\tau \end{bmatrix} \quad . \tag{7}$$

Actually, the evidence for the third family is as yet incomplete (the τ quark, of course, has not yet been discovered and the evidence for ν_τ is indirect). These three families have identical SU(3) x SU(2) x U(1) quantum numbers.

Unfortunately, the SU(5) model does not explain why there are three families. As we will see, any one family furnishes a (reducible) representation of SU(5). We may therefore concentrate on just one family, for instance the first one.

Focusing then on the first family, let us list all of the particle states in the family which have, say, negative or left-handed helicity. Including particles as well as antiparticles, and listing all the color components of quarks and antiquarks, a simple counting shows that there are fifteen states in all:

$$\begin{bmatrix} u \\ u \\ u \end{bmatrix}_L , \quad \begin{bmatrix} d \\ d \\ d \end{bmatrix}_L , \quad \begin{bmatrix} \bar{u} \\ \bar{u} \\ \bar{u} \end{bmatrix}_L , \quad \begin{bmatrix} \bar{d} \\ \bar{d} \\ \bar{d} \end{bmatrix}_L \tag{8}$$

$$\begin{bmatrix} e^- \\ \nu_e \end{bmatrix}_L \quad \text{and} \quad e^+{}_L \quad .$$

I have listed just the states of one helicity, because, as SU(5) is an internal symmetry, an SU(5) operation leaves the helicity unchanged, and the states just of one helicity must by themselves furnish a representation of SU(5). Assuming that only the states indicated in (8) are relevant, we must arrange these 15 states into a 15 dimensional representation of SU(5).

Only a few representations of SU(5) are small enough to be relevant; obviously, any representation with more than 15 elements is not of interest.

The smallest non-trivial representation of SU(5) has five elements. Indeed, SU(5) is <u>defined</u> as the group of 5x5 unitary matrices (of determinant one). The five elements A^i on which these elements act form a five dimensional representation of SU(5), usually referred to simply as the 5. SU(5) has also a second five dimensional representation. Let $B_j = (A^j)^*$ be the complex conjugates of the A^j. The B_j transform according to a second five dimensional representation of SU(5), usually referred to as the $\bar{5}$. It is very useful to distinguish the $\bar{5}$ from the 5 by writing the index down instead of up, B_j as opposed to A^i. From B_j and A^i we may form the quantity $B_i A^i$ (Einstein summation convention is always used and $i=1,\ldots,5$), which is invariant under SU(5) transformations, because of the definition of SU(5) as a group of <u>unitary</u> matrices.

All representations of SU(5) can be formed by using the 5 and $\bar{5}$ as building blocks. Given two objects A^i and $\tilde{A}^j$ that both transform as 5's, we may form the second rank antisymmetric tensor $A^{ij} = A^i\tilde{A}^j - A^j\tilde{A}^i$. It has $5 \cdot 4/2 = 10$ independent components, and transforms according to the next smallest representation of SU(5), usually referred to as the 10. We may again form a new representation, the $\overline{10}$, by complex conjugation: $B_{ij} = (A^{ij})^*$.

We may instead combine two 5's <u>symmetrically</u> to form a symmetric second rank tensor $S^{ij} = A^{i}\tilde{A}^{j} + A^{j}\tilde{A}^{i}$. How many independent components does the symmetric second rank tensor have? With i and j running from 1 to 5, there would be $5^2 = 25$ independent components if no symmetry were imposed. We have already noted that the antisymmetric tensor has 10 components, so the symmetric tensor has $25 - 10 = 15$ independent components. In addition to this representation, denoted as the 15, we can form once again the complex conjugate representation, the $\overline{15}$.

All other representations of SU(5) have more than fifteen elements, so the fifteen states of a family must be assigned somehow to the 5, $\overline{5}$, 10, $\overline{10}$, 15, and $\overline{15}$ representations*. There are essentially five ways one may distribute fifteen states among those six representations:

(i) 15 ,

(ii) 5 + 5 + 5 ,

(iii) 5 + 5 + $\overline{5}$,

(iv) 5 + 10 ,

(v) $\overline{5}$ + 10 .

In addition, there are five other possibilities that differ from the above by replacing all representations by their complex conjugates. Replacing all representations by their complex conjugates is the operation of charge conjugation. The five possibilities that differ from the above only by overall charge conjugation are not essentially different and need not be considered separately.

Of the above five possibilities, the first one is certainly the most attractive. It is the only choice in which all fifteen states

---

* Of course, SU(5) also has a trivial one dimensional representation, on which each group element is represented by the identity. This is the 1, or singlet, representation. A singlet would be annhilated by all SU(3) x SU(2) x U(1) generators (and by all SU(5) generators) so it would not participate in the strong, weak, or electromagnetic interactions. As all fifteen states in a family have at least some of those interactions, the singlet is not relevant in assigning the family to SU(5) multiplets. However, the right-handed neutrino, if it exists, would be an SU(3) x SU(2) x U(1) and SU(5) singlet, and if it is assumed to exist, a new model, the O(10) model, can be constructed.

of a family form a single, irreducible representation of SU(5). However, it is not difficult to show that, of the five possibilities, only the last one is phenomenologically acceptable. This is so far two reasons. Of the above possibilities, only the $\bar{5}$ + 10 is free from Adler-Bell-Jackiw anomalies, which, if present, cause the theory to be mathematically inconsistent. And only the $\bar{5}$ + 10 has the right SU(3) x SU(2) x U(1) quantum numbers to describe a family. I will leave it to you to work out what is wrong with options (i) to (iv) above, but let us now work out the detailed assignment of the states of a family to the $\bar{5}$ + 10.

From the way we have embedded SU(3) x SU(2) in SU(5), the 5 of SU(5) transforms under SU(5)

$$\left[\begin{array}{c} 3 \\ \hline 2 \end{array}\right] \begin{array}{l} \leftarrow \quad SU(3) \\ \\ \leftarrow \quad SU(2) \quad , \end{array} \tag{9}$$

as (3,1) + (1,2); there is a color triplet, weak singlet and a color singlet, weak doublet. The $\bar{5}$ is the complex conjugate of this, or $(\bar{3},1) + (1,\bar{2})$*. The (1,2) must be identified as the $(e^-,\nu_e)^T$ doublet, because this is the only color singlet, weak doublet in the family. What about the $(\bar{3},1)$? At first it seems that the $(\bar{3},1)$ may be identified as the left-handed $\bar{u}$ or $\bar{d}$ antiquark, both of which are weak singlets. However, as the electric charge Q is a generator of SU(5) ($Q = T_3 + Y/2$), and as all generators are traceless, we must ensure that the sum of the elecric charges of all particles in the $\bar{5}$ vanishes. This works only if we choose the $\bar{d}$; recalling that the $\bar{d}$ has three colors, the sum of charges is then 1/3 + 1/3 + 1/3 − 1 + 0 = 0. We have thus identified the components of the $\bar{5}$:

---

* Under SU(3), quarks and antiquarks transform as 3 and 3 respectively; these are inequivalent, complex conjugate representations. However, under SU(2) there is only one two dimensional representation, the 2. Its complex conjugate transforms in the same way. This is most famous in the case in which SU(2) is identified as ordinary angular momentum. We all know there is only one inducible two dimensional representation, which describes spin one half. In SU(2) there is no essential distinction between a vector ϕ^i or ψ_j with index up or down because the index can be raised or lowered with the invariant antisymmetric second rank tensor ε_{ij}.

$$\begin{bmatrix} \bar{d} \\ \bar{d} \\ \bar{d} \\ \hline e^- \\ \nu_e \end{bmatrix}_L . \tag{10}$$

It is only slightly more difficult to unravel the composition of the 10. As the 10 is the antisymmetric combination of two 5's, and the 5 transforms as (3,1) + (1,2) under SU(3) x SU(2), the 10 transforms as

$$\begin{aligned} [((3,1) + (1,2)) \otimes ((3,1) + (1,2))]_A &= [(3,1) \otimes (3,1)]_A \\ &+ (3,1) \otimes (1,2) + [(1,2) \otimes (1,2)]_A \ , \end{aligned} \tag{11}$$

where the subscript A denotes the antisymmetric product. Now, in SU(3), the antisymmetric product of two triplets transforms as an antitriplet. And in SU(2), the antisymmetric product of two doublets (spin or isospin one half) is a singlet (spin or isospin zero). So the 10 transforms as

$$(\bar{3},1) + (3,2) + (1,1) \quad . \tag{12}$$

Here the $(\bar{3},1)$ must be $\bar{u}_L$, the only antiquark not already assigned. The (3, 2) must be $(u,d)_L{}^T$ which, indeed, make up a color triplet, weak doublet. And the (1,1) must be $e^+{}_L$, the only color singlet, weak singlet in the family. So the quantum numbers of those members of the family not already assigned to the $\bar{5}$ just make up the 10.

The 10, being an antisymmetric, second rank tensor, is conveniently arranged as an antisymmetric matrix:

$$\left[\begin{array}{ccc|cc} 0 & \bar{u} & \bar{u} & u & d \\ -\bar{u} & 0 & \bar{u} & u & d \\ -\bar{u} & -\bar{u} & 0 & u & d \\ \hline -u & -u & -u & 0 & e^+ \\ -d & -d & -d & -e^+ & 0 \end{array}\right]_L . \tag{13}$$

Perhaps the most important aspect of this is that we have placed quarks, antiquarks and leptons in the same representation of

SU(5). This leads to proton decay. In fact, the gauge bosons of SU(5) that are <u>not</u> in SU(3) x SU(2) x U(1)

$$\left[\begin{array}{ccc|cc} & & & X & Y \\ & \bigcirc & & X & Y \\ & & & X & Y \\ \hline X^* & X^* & X^* & & \\ Y^* & Y^* & Y^* & \multicolumn{2}{c}{\bigcirc} \end{array}\right], \qquad (14)$$

mediate processes such as uu → X → $\bar{d}$ e^+ (figure 1). Adding a spectator d quark, this is uud → $e^+ d\bar{d}$, which can represent the decay $p \to e^+\pi^0$. The fact that quark-lepton unification can lead to proton decay was first suggested by Pati and Salam[2] , in a somewhat different context.

This brings us to the subject of symmetry breaking. In fact, the SU(5) gauge theory describes color forces, which are very strong, the weak and electromagnetic interactions, which are far weaker, and proton decay, which is very, very weak. The SU(5) symmetry must be spontaneously broken to account for the different strenghts of these interactions.

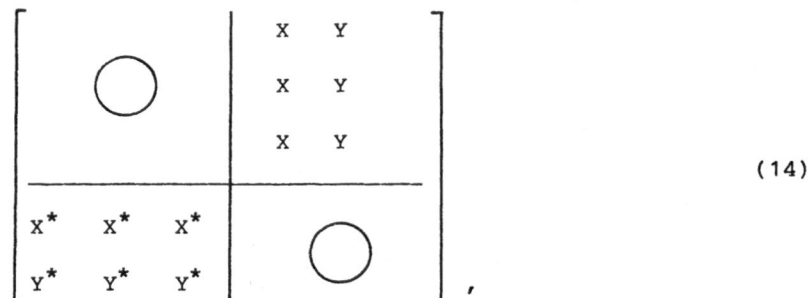

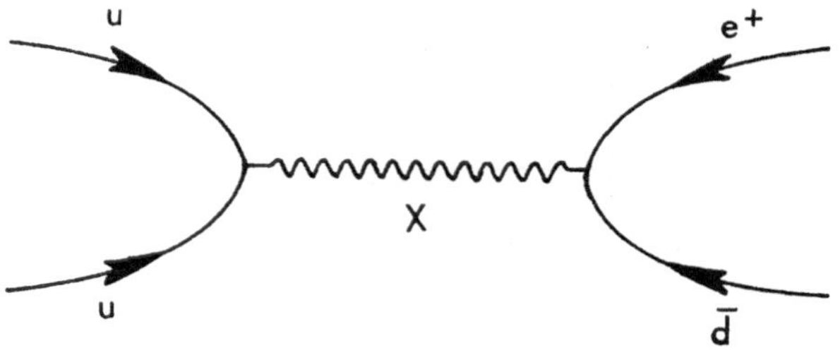

Fig. 1. Graph for the process uu → X → d e^+

In the first stage of symmetry breaking, we spontaneously break SU(5) down to the gauge symmetry SU(3) x SU(2) x U(1) of the strong, weak, and electromagnetic interactions. This is accomplished by means of the vacuum expectation value of a field $\phi^i{}_j$ that transforms in the adjoint representation of SU(5). In other words, $\phi^i{}_j$ is a traceless, hermitian, 5x5 matrix.

Since we ultimately wish to discuss supersymmetry, I might pause at this moment to note that since the gauge bosons $A_\mu{}^i{}_j$ also transform in the adjoint representation of SU(5), one might hope that supersymmetry would relate ϕ to A_μ. Perhaps supersymmetry, combined with SU(5), would predict the existence of ϕ. Unfortunately, like many other attempts to make use of supersymmetry from a phenomenological point of view, this runs into difficulties. The simple (N=1) supersymmetries relate gauge bosons only to particles of spin 1/2. To relate gauge bosons to particles of spin 0, one needs extended supersymmetry (N $\geq$ 2). But in extended supersymmetry it can be readily shown that the spin 1/2 fermions necessarily transform in a real representation of the gauge group; this contradicts the observed properties of quarks and leptons.

Having introduced the field ϕ, one next chooses the Lagrangian so that the expectation value of ϕ has the form

$$\langle \phi^i{}_j \rangle = M \begin{bmatrix} 2 & & & & \\ & 2 & & & \\ & & 2 & & \\ & & & -3 & \\ & & & & -3 \end{bmatrix} . \tag{15}$$

This vacuum expectation value spontaneously breaks SU(5) down to the subgroup consisting of matrices that commute with $\phi^i{}_j$. That subgroup is precisely SU(3) x SU(2) x U(1), so we have accomplished our goal of spontaneously breaking SU(5) down to the observed gauge group of the strong, weak, and electromagnetic interactions.

This symmetry breaking gives masses of order M to the gauge bosons X and Y that mediate proton decay. The proton lifetime is therefore proportional to M^4. To ensure an aceeptably long lifetime for the proton, M must be about 10^{15} GeV or longer.

After spontaneously breaking SU(5) down to SU(3) x SU(2) x U(1) we still must worry about the spontaneous breaking of SU(2) x SU(1) down to the U(1) gauge group of electromagnetism. In the Weinberg-Salam model, the spontaneous breaking of SU(2) x U(1) is accomplished by means of the vacuum expectation value of a Higgs doublet. We now must ask what SU(5) representation contains particles with the quantum numbers of the usual Weinberg-Salam doublet. This question is easily answered. In fact, from our previous discussion it follows that the fundamental 5 representation of SU(5) is suitable. We introduce a second Higgs multiplet h^i transforming in this representation. The lower two components

$$\left[\begin{array}{c} h^1 \\ h^2 \\ h^3 \\ \hline h^+ \\ h^0 \end{array}\right] , \tag{16}$$

of h^i are a color singlet, SU(2) doublet which can be interpreted as the conventional Higgs multiplet. We therefore can obtain SU(2) x U(1) breaking from the expectation value $\langle h^0 \rangle$ of the neutral Higgs boson. Since the Fermi constant of weak interactions is $1/\langle h^0 \rangle^2$, we must have $\langle h^0 \rangle$ approximately 250 GeV.

We now come to the question of quark and lepton masses. As we have seen, the left-handed quarks and leptons are to be placed in the 10 and $\overline{5}$ of SU(5); we may denote these as $\psi_L{}^{ij}$ and $\tilde{\psi}_{Lk}$, respectively. As in the original Weinberg-Salam model, gauge invariance forbids bare masses for the quarks and leptons. The quarks and leptons must instead get masses from their coupling to the Higgs boson that breaks SU(2) x U(1) symmetry. There are two possible couplings. One may couple h^i to two ψ's

$$\lambda \, \varepsilon_{ijk\ell m} \, h^i \, \psi_L{}^{jk} \, \psi_L{}^{\ell m} \tag{17}$$

by making use of the invariant antisymmetric tensor of SU(5), $\varepsilon_{ijk\ell m}$. One also may couple the complex conjugate of h to a ψ and a ψ:

$$\tilde{\lambda} \, h_i{}^* \, \psi_L{}^{ij} \, \tilde{\psi}_{Lj} \tag{18}$$

The two coupling constants λ and $\tilde{\lambda}$ are arbitrary. Expressions (17) and (18) are the only possible couplings of h to left-handed quarks; the couplings to right-handed quarks are just the hermitian conjugates of (17) and (18).

As (17) and (18) are the only possible couplings, they must account for all masses within a single generation of quarks and leptons. This means that the up quark, down quark, and electron masses must be accounted for. (The neutrino masses are zero in this model; but see below.) Since three masses (u, d, and e) are determined in this model by only two arbitrary coupling constants, λ and $\tilde{\lambda}$, the model obviously predicts one relationship among these three masses. Let us determine what this relation is.

Since $\langle h^i \rangle = \delta^{i5} \langle h^0 \rangle$, the relevant portion of (17) is

$$8\lambda\langle h^0\rangle \left(\psi_L{}^{12}\,\psi_L{}^{34} + \psi_L{}^{23}\,\psi_L{}^{41} + \psi_L{}^{34}\,\psi_L{}^{12}\right) \quad . \qquad (19)$$

Looking back at our detailed identification of the components of $\psi_L{}^{ij}$ (equation (13)), we see that the three terms in (19) give the arbitrary mass $8\lambda\langle h^0\rangle$ to the three colors of up quarks. Instead, the relevant portion of (18) is

$$\tilde{\lambda}\langle h^0\rangle \left(\psi_L{}^{15}\,\tilde{\psi}_{L1} + \psi_L{}^{25}\,\tilde{\psi}_{L2} + \psi_L{}^{35}\,\tilde{\psi}_{L3} + \psi_L{}^{45}\,\tilde{\psi}_{L4}\right) \quad . \qquad (20)$$

Looking back again at the explicit identifications made earlier, we see that the first three terms in (20) give the mass $\tilde{\lambda}\langle h^0\rangle$ to the three colors of d quarks, and the last term gives the same mass to the electron.

The prediction of the SU(5) model is thus $m_d = m_e$. Actually, this prediction holds only at the tree level. Because of symmetry breaking, the higher order corrections affect m_d and m_e differently. After renormalization, the SU(5) prediction turns out to be[3] approximately $m_d = 3m_e$. The SU(5) model makes for the other families the analogous predictions $m_s = 3m_\mu$ and $m_b = 3m_\tau$. Unfortunately, of these predictions, only the last one is reasonably successful.

Let us now summarize the basic features of this model.

(1) Because the strong, weak, and electromagnetic interactions are all part of SU(5), the model predicts relations among the coupling constants of these interactions. In the case of the weak and electromagnetic interactions, the relationship can be expressed in the form $\sin^2\theta_W = 0.21$.[4] This is in excellent agreement with experiment.

(2) Because of symmetry breaking effects, the model does not really predict the value of the strong coupling constant. The prediction depends on the a priori unknown value of the mass M at which SU(5) is broken. It is more useful to take the observed value of the strong coupling (or more pertinently, the observed value of the energy scale at which the strong interactions become strong), and infer from this what M must be. One finds that M is around 10^{15} GeV (in the simplest form of the SU(5) model). Once M is known, the model predicts the proton lifetime. The predicted lifetime is around 10^{30} years. Of course, this prediction has not yet been confirmed or refuted.

(3) As we have just seen, the model predicts the mass relations $m_d = 3m_e$, $m_s = 3m_\mu$, $m_b = 3m_\tau$. Of these, only the third is successful. There are many ways to modify these predictions. One may add new Higgs bosons which make new contributions to the quark and lepton masses. One may add new fermions which, by mixing with the

usual quarks and leptons, modify the predicted mass relations. I do not believe that a convincing resolution of this problem has yet emerged.

(4) The simplest form of the SU(5) model predicts that the neutrino masses vanish. This is related to the fact that while in this model baryon number is not conserved, and lepton number is not conserved, the difference B-L is conserved. In many interesting modifications of this model (with a larger gauge group or additional fields of spin 0 or spin 1/2), the neutrino masses are predicted to be small but non-zero.

(5) Lest we forget, the quantum numbers of the quarks and leptons are just right to make the SU(5) model work! I believe that this is actually the best reason to suspect that it has some truth in it.

The most serious problem with this picture is probably the "gauge hierarchy problem"[5]. As we have seen, the mass scale M of SU(5) breaking is about 10^{15} GeV, in this model. The mass scale $\langle h^0 \rangle$ of SU(2) x U(1) breaking is about 250 GeV. The ratio $M/\langle h^0 \rangle$ is larger than 10^{12}. Ever since the work of Dirac[6] on the "large numbers", it has been widely felt that such unreasonably large numbers should not be arbitrarily postulated, but should have some definite explanation. The SU(5) model, in its present form, does not give any explanation for the enormously large ratio $M/\langle h^0 \rangle$. This ratio is obtained by an arbitrary adjustment or "fine tuning" of parameters.

II. Supersymmetric Models

Let us now discuss some attempts at supersymmetric generalizations of the grand unified models that we considered yesterday.

There are several possible motivations for exploring the supersymmetric generalizations of our current theories.

First of all, the current theories are rather untidy in certain respects. In constructing these theories one first of all chooses a gauge group and a fermion representation. Assuming that (as in the most interesting cases) the fermion representation is such that bare masses are impossible, the only free parameter that arises here is the gauge coupling constant, a single constant, in the case of a unified gauge group.

Thus, this part of the theory, which includes the most directly observed interactions, is relatively free of adjustable parameters. But to make the theory work, it must be backed up by a rich superstructure, the Higgs bosons, with their self-couplings (and vacuum expectation values) that drive symmetry breaking, and their Yukawa interactions that are responsible for fermion masses. Associated with the Higgs sector there are many freely adjustable coupling constants, about fifteen in the SU(3) x SU(2) x U(1) model and even more in SU(5). We have at present no physical principle that determines the quantum numbers and couplings of Higgs bosons. This certainly is one of the main deficiencies in the present theories. A very worthy goal in constructing a supersymmetric theory would be to try to use supersymmetry to cut down on the arbitrariness associated with Higgs bosons.

Unfortunately, this goal is rather elusive, for a variety of reasons, some of which we will come to later.

Another problem to which supersymmetry might plausibly be applied is the gauge hierarchy problem, which I explained briefly at the end of the last lecture. This is again a question about Higgs bosons, but now the problem is to explain the great disparity between the mass scale of SU(5) breaking and the scale of SU(2) x SU(1) breaking. The problem is equivalent to explaining why the Higgs boson of the SU(2) x U(1) model is so light compared to the mass scale of the SU(5) model. The basic reason to suspect that supersymmetry might be relevant to this problem is that it is the only known symmetry that can restrict the masses of spin zero particles.

Several alternative ways to apply supersymmetry to the gauge hierarchy problem have been considered. As I have discussed the matter in great detail elsewhere[7,8] I will not pursue it further in these lecture notes.

I actually believe that the best reason to think seriouly about possible applications of supersymmetry in particle physics is simply the fact that sooner or later, if we are to make progress in particle physics, we will have to incorporate new ideas. Supersymmetry is one of the best candidates as a new idea that should be incorporated in our understanding of particle physics.

In any case, without worrying too much about specific motivations, let us discuss what happens when one tries to extend the SU(5) model, which we discussed yesterday, to include supersymmetry.

At the outset, we must decide at what energy scale we think supersymmetry is spontaneously broken. Supersymmetry certainly is spontaneously broken, if it is relevant to nature at all, because we do not observe in nature bosons and fermions of the same mass. The SU(5) model has, basically, two mass scales, the large scale at which SU(5) is broken to SU(3) x SU(2) x U(1) and the small scale at which SU(2) x U(1) is broken to U(1). Do we wish to assume that supersymmetry is spontaneously broken at the large mass scale of the SU(5) model, at the small scale, or at another energy scale altogether?

In common with much recent work, we will assume that supersymmetry is spontaneously broken only at relatively low energies, at the energy scale of the SU(2) x U(1) model. This is the strongest assumption that can be made, in the sense that it leads to the most severe restrictions on the theory at ordinary energies. However, we should bear in mind that supersymmetry may be relevant to nature but spontaneously broken at very high energies, perhaps even at the Planck mass. In this case, the implications of supersymmetry to particle physics at ordinary energies would be more indirect. Ordinary particle physics would not be described by an ordinary, renormalizable, supersymmetric Lagrangian; but supersymmetry would presumably still place strong restrictions on the masses, quantum numbers, and interactions of the particles surviving to low energies.

Let us now turn to the nitty-gritty of supersymmetric theories. Supersymmetric theories with particles of spin 0 and spin 1/2 only are readily described as follows. We introduce an arbitrary number of complex scalar fields A^i, i=1...n. Each A^i has a supersymmetric partner, which is a left-handed spin one half field $\chi_L{}^i$. The supersymmetric partners of the complex conjugates A_j^* of the A^i are the right-handed spinor fields $\chi_{Rj} = (\chi_L{}^j)^*$. It is often convenient to group A^i and χ^i as a supermultiplet $(A^i, \chi_L{}^i)^T$.

In addition, we introduce a function $W(A^i)$, which is known as the "superspace potential". W must be an analytic function of the A^i, in other words, a function of the A^i but not of their complex conjugates $A_j{}^*$.

A supersymmetric Lagrangian can then be written in the form

$$L = L_{kin} + L_{scalar} + L_{Yukawa} \quad . \tag{21}$$

Here L_{kin} is the standard kinetic energy

$$L_{kin} = \partial_\mu A_k^* \partial_\mu A^k + \overline{\chi_k} i \not{\partial} \chi^k \quad , \tag{22}$$

while L_{scalar} is the scalar potential

$$L_{scalar} = -V(A^i, A_j{}^*) = -\sum_i \left| \frac{\partial W}{\partial A^i} \right|^2 \quad , \tag{23}$$

and L_{Yukawa} describes the Yukawa interactions

$$L_{Yukawa} = \frac{\partial^2 W}{\partial A^i \, \partial A^j} \chi_L{}^i \chi_L{}^i + \text{h.c.} \quad . \tag{24}$$

How this sort of theory is constructed (and shown to be supersymmetric) has been discussed at this school by other lecturers, and I will not go into details.

Depending on the choice of the superspace potential W, this Lagrangian may have global symmetries. Any such symmetries, if anomaly free, can be gauged. One introduces gauge fields $A_\mu{}^a$ and their supersymmetric partners $\lambda_\alpha{}^a$. One adds to the Lagrangian the usual kinetic energy of $A_\mu{}^a$ and $\lambda_\alpha{}^a$,

$$\Delta L_{kin} = -1/4\ (F_{\mu\nu}{}^a)^2 + \overline{\lambda}^a\ i\not{D}\lambda^a \quad . \tag{25}$$

One replaces derivatives by covariant derivates, $\partial_\mu \to D_\mu$, in (22) above. Also, supersymmetry requires the addition of an extra term to the potential energy,

$$\Delta L_{scalar} = -\Delta V = -1/2 \sum_a e_a{}^2 \left|(A^*,\ T^a A)\right|^2 \quad . \tag{26}$$

Here the sum runs over all generators a of the gauge group, the e_a are the gauge coupling constants, and T^a are the generators of the gauge group, acting on the representation of the group furnished by the A^i. Finally, supersymmetry also requires an extra Yukawa coupling

$$\Delta L_{Yuk} = \sum_a e_a\ \lambda_{L\alpha}{}^a\ (A^*,\ T^a\ \chi_L{}^\alpha) \quad , \tag{27}$$

which will not figure in our discussions.

The most convenient way to develop and understand these formulas is the superspace formulation, which has been developed at this school by Grisaru. What I have described above is the most general renormalizable supersymmetric theory, except that if the gauge group contains a U(1) factor, it is possible to add to the Lagrangian an extra coupling, the so-called Fayet-Iliopoulos D term.

For many purposes, the most important formula is the formula for the scalar potential,

$$V(A^i,\ A_j{}^*) = \sum_i \left|\frac{\partial W}{\partial A^i}\right|^2 + 1/2 \sum_a e_a{}^2 \left|(A^*,\ T^a A)\right|^2 \quad . \tag{28}$$

By minimizing this potential, one determines, in the simplest weak coupling approximation, what the vacuum state is. One can determine, in this approximation, what gauge symmetries are spontaneously broken. One can also determine whether supersymmetry is spontaneously broken.

As is the case for any symmetry, supersymmetry is spontaneously broken if the symmetry generators Q_α do not annihilate the vacuum state $|\Omega\rangle$. If $Q_\alpha |\Omega\rangle \neq 0$, supersymmetry is spontaneously broken; bosons and fermions are not degenerate despite the underlying supersymmetry. This condition can be conveniently re-expressed, by using the basic relation that in supersymmetric theories the Hamiltonian H is the sum of the squares of the supersymmetry charges

$$H = 1/2 \sum_\alpha (Q_\alpha)^2 \quad . \tag{29}$$

Here the Q_α are hermitian, so (29) implies that $Q_\alpha|\Omega\rangle = 0$ if and only if $H|\Omega\rangle = 0$. We conclude that supersymmetry is unbroken if and only if the energy of the ground state is zero.

In the classical approximation, the zero point energy of the fields may be neglected, and the energy of the ground state just equals the minimum of the potential $V(A^i, A^*_j)$. Supersymmetry is unbroken if V = 0 for some value of the A^i (which is then their vacuum expectation value, in this simplest approximation). As V is a sum of squares, V = 0 only if each term separately vanishes. Thus, the condition for the energy to vanish in the classical limit is that

$$\frac{\partial W}{\partial A^i} = 0 \quad , \tag{30}$$

for each field A^i and that

$$(A^*, T^a A) = 0 \quad , \tag{31}$$

for every generator T^a of the gauge group.

Now, in discussing supersymmetric grand unified theories, we will consider theories in which, in perturbation theory, the unified group is spontaneously broken to SU(3) x SU(2) x U(1) but supersymmetry is <u>unbroken</u>. We will assume that SU(2) x U(1) breaking and supersymmetry breaking occur at much lower energies because of some non-perturbative symmetry breaking mechanism. Theories in which supersymmetry is spontaneously broken at the tree level have very interesting properties, some of which I have discussed elsewhere[8] but we will not discus this option here.

Restricting ourselves to SU(5) models, the first step is to describe, in the context of supersymmetry, the breaking of SU(5) to SU(3) x SU(2) x U(1). As we discussed yesterday, the first stage of symmetry breaking is usually achieved via the vacuum expectation value of a field transforming in the adjoint representaion of SU(5). So here we introduce a <u>supermultiplet</u> consisting of a 5x5 traceless, complex matrix $C^i{}_j$ and its fermi partner $\chi^i{}_j$. Notice that, while an irreducible representation of SU(5) would consist of a traceless hermitian matrix, supersymmetry does not permit us to require that C be hermitian. After all, the basic ingredient in supersymmetry is the complex scalar field.

We now must choose the superspace potential W. W must be a function of C only, and not C^*, and for a renormalizable theory W must be at most cubic. The most general choice compatible with SU(5) symmetry is (apart from an irrelevant additive constant)

$$W(C) = \frac{\lambda}{3} \mathrm{Tr}\, C^3 + \frac{M}{2} \mathrm{Tr}\, C^2 \quad , \tag{32}$$

where M and λ are constants.

We now would like to determine for what values of C, if any, the energy vanishes in the classical approximation. Equation (31), in the case of a field in the adjoint representation, can be rewritten in the much more transparent form

$$\left[C^*, C\right] = 0 \tag{33}$$

The statement that C commutes with its adjoint means that C can be diagonalized by an SU(5) transformation, which would not be true, a priori, for a general 5x5 complex matrix. Equation (30) then gives

$$\lambda\left((C^2)^i{}_j - 1/5\ \delta^i{}_j\ \mathrm{Tr}\, C^2\right) + M\, C^i{}_j = 0 \quad . \tag{34}$$

Note that, in calculating $\partial W/\partial C^i{}_j$, one must respect the condition that $C^i{}_j$ is traceless.

Armed with the knowledge that C can be diagonalized, it is not difficult to show that (34) has essentially three solutions. Up to a gauge transformation, we have

$$C^i{}_j = 0 \quad ,$$

$$C^i{}_j = \frac{M}{3\lambda} \begin{bmatrix} -4 & & & & \\ & 1 & & & \\ & & 1 & & \\ & & & 1 & \\ & & & & 1 \end{bmatrix} \quad , \tag{35}$$

$$C^i{}_j = \frac{2M}{\lambda} \begin{bmatrix} -3/2 & & & & \\ & -3/2 & & & \\ & & 1 & & \\ & & & 1 & \\ & & & & 1 \end{bmatrix} \quad .$$

These three solutions correspond, respectively, to the unbroken gauge groups SU(5), SU(4) x U(1), and SU(3) x SU(2) x U(1).

In the classical approximation, these three states are exactly degenerate, at zero energy. They were found by solving the equations (30) and (31) of unbroken supersymmetry. There is no way, in the classical approximation, to decide what is the proper choice of the vacuum expectation value of $C^i{}_j$; it is therefore also impossible to determine what is the unbroken gauge group.

This type of degeneracy is a characteristic feature of supersymmetric theories. One would hardly expect, in most theories, to have a degeneracy between local minima of the potential that are not related by any symmetry. More exactly, one would hardly expect to achieve such a degeneracy except by artificially adjusting the parameters in the potential. But in the supersymmetric case such degeneracies between thoroughly inequivalent states occur quite characteristically.

Even more remarkably, it can be shown that the degeneracy between the three solutions given in equation (35) is not lifted in any finite order of perturbation theory. In all finite orders, they remain at zero energy, and supersymmetry is unbroken.

What happens at the non-perturbative level, in theories of this type, is a very difficult question, although some results can be obtained[9] . We will simply assume, in the remainder of our discussion, that in this theory, or one like it, the non-perturbative corrections spontaneously break supersymmetry, giving a non-zero energy to all candidate vacuum states, but choosing the state with unbroken gauge symmetry SU(3) x SU(2) x U(1) as the true vacuum state, the one with lowest energy.

To make a realistic model, the next step would be to include fields in the 10 and $\overline{5}$ representations of SU(5). As discussed in the first lecture, these are the representations that describe quarks and leptons. So we introduce scalar fields M^{ij} (= $-M^{ji}$) and $\tilde{M}_K$ in the 10 and $\overline{5}$ representations, along with their fermi partners $\psi_L{}^{ij}$ and $\tilde{\psi}_{Lk}$, which are the quarks and leptons.

Actually, since we wish to describe at least three generations, the fields must carry an extra label, the generation label. We may actually write the scalars as $M^{ij}{}_a$ and $\tilde{M}_{kb}$, where a, b = 1...3; a and b distinguish the generations. The fermions likewise carry this new label.

Now, as we have seen in the first lecture, the $\overline{5}$ fields $\tilde{M}_{kb}$ have components that transform like the doublet of the Weinberg-Salam model (see equation (16)). They are simply the supersymmetric partners of the doublets $(\ell^-, \nu_\ell)_L{}^T$ where ℓ^- may be e, μ, or τ.

One's first thought may therefore be that it is the neutral component of one of the $\tilde{M}_{kb}$ whose vacuum expectation value breakes SU(2) x U(1) down to U(1).

If this idea worked, it would be an important achievement of supersymmetry. We would have obtained a rationale for the existence of the Weinberg-Salam Higgs doublet.

Unfortunately, there are many difficulties with this idea. The problems arise when one tries to obtain quark and lepton masses. As we know, in the SU(5) model the fermion masses come from Yukawa couplings of the Higgs boson to quarks and leptons. In our supersymmetric context, there are many difficulties associated with these Yukawa couplings.

The most severe problem pertains to the proton lifetime. However, let us postpone discussion of that and first consider the difficulties that arise in obtaining the right quark and lepton mass spectrum. We must first introduce a superspace potential. The most general cubic, SU(5) invariant form is

$$W(M^{ij}{}_a, \tilde{M}_{kb}) = g_{abc}\, M^{ij}{}_a\, \tilde{M}_{ib}\, \tilde{M}_{jc} \quad . \tag{36}$$

Here the g_{abc} are dimensionless coupling constants. For future use note that $g_{abc} = -g_{acb}$ because $M^{ij}{}_a$ is antisymmetric in i and j.

Now, recalling that the Yukawa couplings have the general form $(\partial^2 W/\partial A^i\, \partial A^j)\, \psi^i\, \psi^j$, we see that (36) gives rise to the Yukawa coupling

$$L_{Yuk} = 2g_{abc}\, \psi^{ij}{}_a\, \tilde{\psi}_{ib}\, \tilde{M}_{jc} \tag{37}$$

This is just the sort of coupling which, in the standard SU(5) model, gives mass to the d quark and to the electron (compare with equation (18)). If one of the $\tilde{M}_{jc}$ does have a vacuum expectation value, (37) will give mass to the down quarks and charged leptons (except for a proviso that will be stated shortly).

However, we do not get masses for up quarks! The supersymmetric model, unless extra fields are introduced, simply does not allow a generalization of the coupling (17) which, in the simplest SU(5) model, gives masses to up quarks. Such a coupling would come from a term in W of the type $\Delta W \sim \varepsilon_{ijk\ell m} M^{ij} M^{k\ell} \tilde{M}^{*m}$. This term is forbidden; because of the dependence on $\tilde{M}^*$ as well as M, it violates the requirement that W should be an analytic function of M and $\tilde{M}$. We thus arrive at the uncomfortable conclusion that this model predicts $m_u = m_c = m_t = 0$.

Actually, a closer thought shows that at the tree level this model also predicts that there is one massless down quark and one massless charged lepton. Indeed, because of the antisymmetry $g_{abc} = -g_{acb}$, (37) does not give mass to the down quark and lepton in the same generation as the Higgs field $\tilde{M}_{kb}$ that has an expectation value. Actually, the prediction that $m_d = m_e = 0$ at the tree level is attractive. The down quark and electron are so light that it would be desirable for them to be massless at the tree level as long as the model is arranged so that they do in fact get masses from the higher order corrections. That is presumably the case here (since there is no chiral symmetry that would keep $m_d = m_e = 0$), although the details, which would merit further examination, depend on what assumption one makes about the mechanism for supersymmetry breaking. What is really unacceptable about the relation $m_u = m_c = m_t = 0$ that we found a moment ago is that it will be preserved in higher orders. It is enforced in higher orders because of certain chiral symmetries that hold because some of the usual SU(5) Yukawa couplings are missing.

Even worse, the introduction of the superspace potential (36) leads to catastrophically fast proton decay. Along with the desired coupling (37) that gives masses to some quarks and leptons, (36) leads to other couplings that violate baryon number, couplings of the color triplet components of $\tilde{M}$ to two quarks or to an antiquark and a lepton.

Even without supersymmetry, SU(5) relates the needed couplings of the SU(2) x U(1) doublet to baryon number violating couplings of their color triplet partners. However, without supersymmetry, one is free to assume that the SU(2) x U(1) doublet is light, but its color triplet partners are extremely heavy. It is not understood why this would be so; to explain it is an important problem, part of the gauge hierarchy problem. But, without supersymmetry, one is free to assume that the color triplets are vastly heavier than the SU(2) x U(1) doublet, and to maintain this state of affairs via an arbitrary adjustment of parameters.

With supersymmetry, we are in a more difficult situation. We have assumed that the Higgs doublet is the supersymmetric partner of one of the lepton doublets. This means that the color triplet partners of the Higgs are related by supersymmetry to some of the ordinary quarks and antiquarks. If supersymmetry is broken only at ordinary particle physics energies, the supersymmetric partners of ordinary quarks and antiquarks cannot be vastly heavier than the quarks and antiquarks themselves. Hence, if supersymmetry survives down to relatively low energies, (36) leads to catastrophically fast proton decay, the lifetime being a small fraction of a second.

We are, of course, free to assume that supersymmetry is broken at extremely high energies, the energy of grand unification or above. In this case, the Higgs doublet could remain light while the color triplets become heavy, although, just as in conventional

SU(5), it is not at all clear how this would come about in a natural way. However, if we hope to solve the gauge hierarchy problem by using an unbroken supersymmetry to keep the Higgs doublet light down to low energies, the scalar partners of quarks cannot be vastly ligher than quarks themselves.

Alternatively, we could abandon the coupling (36). Then all quarks and leptons are massless, hardly an acceptable state of affairs.

We reluctantly conclude that the idea of identifying the Higgs doublet with the scalar partners of leptons must be abandoned, at least in the framework considered here. Instead, we introduce additional SU(5) multiplets to accommodate the Higgs doublet. The simplest approach is to introduce two new supermultiplets, a 5 $(E^i, \eta^i{}_L)^T$ and a $\overline{5}$ $(F_j, \eta_{jL})^T$. We will use E^i and F_j as Higgs bosons. We introduce both E^i and F_j for two reasons. The 5 and $\overline{5}$ have equal and opposite Adler-Bell-Jackiw anomalies, which cancel when they are introduced together. In addition, to avoid problems like the relation $m_u = m_c = m_t = 0$ that was found above, we need both the 5 and $\overline{5}$.

From the superspace potential we now omit the self-coupling of quarks considered earlier. It leads to too much trouble with the proton lifetime. We try instead

$$\Delta W = \varepsilon_{ijk\ell m} E^i M^{jk}{}_a M^{\ell m}{}_b g_{ab} + F_i M^{ij}{}_a \tilde{M}_{jb} \tilde{g}_{ab} \quad . \qquad (38)$$

Assuming that the neutral components of E and F have vacuum expectation values, (38) gives masses to all quarks and charged leptons. These masses are arbitrary except that the SU(5) relations $m_d = m_e$, $m_s = m_\mu$, $m_b = m_\tau$ are respected. Note that the coupling constants g_{ab} and $\tilde{g}_{ab}$ in (38) are directly related to the up and down quark mass matrices.

We still have the problem that the color triplet components of E and F may mediate proton decay. However, as these particles are now not assumed to be related by supersymmetry to any known fermions, we may suppose them to be extremely heavy even if we believe that supersymmetry is broken only al low energies.

Because E and F together form a real representation of SU(5), they can have bare masses. They also can receive mass by coupling to the field $C^i{}_j$ in the adjoint representation of SU(5) that was introduced at the beginning of this discussion (the field whose vacuum expectation value breaks SU(5) down to SU(3) x SU(2) x U(1). Thus, we add to W an a new term

$$\Delta W = \tilde{M} E^i F_i + \lambda E^i F_j C^j{}_i \qquad (39)$$

By suitable adjustment of the mass $\tilde{M}$ and the coupling λ, one may ensure that the doublet components of E and F are light, but their color triplet partners are extremely heavy. The couplings of the doublets to quarks and leptons could thus give masses to those particles, without causing a rapid proton decay.

The adjustment of $\tilde{M}$ and λ that is needed here is unsatisfactory like the adjustment that is needed in the conventional SU(5) model. There is one advantage in the supersymmetric case, but I do not think it is decisive. The advantage is that, because of the "non-renormalization" theorems of supersymmetric theories, the adjustment of parameters that makes the doublet much lighter than the triplet, if made at the tree level, need not be corrected in higher orders of perturbation theory.

Recently, Dimopoulos and Wilczek[10] have proposed a class of models in which, for group theoretical reasons, the doublet remains massless at the tree level while its color triplet partners gain (large) masses. Their approach involves going to gauge groups larger than SU(5). The approach is interesting, although a realistic model hasn't yet emerged.

The superspace potential presented above (equations (32), (38), and (39)) describes a theory that is technically natural, certain global symmetries forbid the terms that have been omitted. It is, however, an incomplete theory. Although the doublet components of E and F may be light, we have not arranged for any terms in the classical potential that would force the doublets to have vacuum expectation values. Nor have we provided for supersymmetry breaking. One may add extra fields and extra interactions to obtain the desired symmetry breaking. However, within the context of the model described here, it would be more attractive to ask whether the breaking of supersymmetry and SU(2) x U(1) could occur automatically dynamically, because of very small (non-perturbative) effects that could become important at very low energies. In some respects this appears very plausible;[7] on the other hand, there are no positive results, and it is possible to derive some "no-go" theorems[9].

In conclusion, I wish to summarize the present status of a few phenomenological issues in theories along the general lines discussed above. For more detail I refer you to some recent articles[11].

(1) Proton lifetime. Because of the contributions to renormalization of the supersymmetric partners of gauge bosons, the unification mass is several orders of magnitude bigger in supersymmetric grand unified theories than in the conventional theories[12]. One might expect this to dramatically increase the proton lifetime. However, as pointed out by Weinberg[13], in supersymmetric theories proton decay may be mediated more rapidly by the partners of gauge bosons than by the gauge bosons themselves. The present status is

not absolutely clear, but it appears that the proton lifetime is about the same in supersymmetric SU(5) as in conventional SU(5), but the branching ratios are dramatically different.

(2) $\sin^2\theta_W$. Supersymmetric SU(5) predicts a slightly higher value of $\sin^2\theta_W$ than is predicted in the absence of supersymmetry. This worsens the excellent agreement with experiment of the conventional SU(5) theory.

(3) Mass relations. The renormalization effects that correct the naive mass relations and lead to the predictions $m_d = 3m_e$, $m_s = 3m_\mu$, $m_b = 3m_\tau$ are substantially unmodified by incorporation of supersymmetry.

Finally, I should comment on one last phenomenological issue. In supersymmetric theories in which the low energy gauge group is SU(3) x SU(2) x U(1), there are certain difficulties at the tree level with the mass spectrum. One cannot give suitably large masses to the scalar partners of quarks and leptons (which should weight at least 10 or 20 GeV, to explain the failure to observe them so far). One way out is to enlarge the low energy gauge group. Fayet[14] showed that the scalars can receive appropriate masses in a model with a low energy gauge group SU(3) x SU(2) x U(1) x U(1). This approach has recently been developed and extended by Weinberg[13].

Another approach is to relay on loop diagrams to give masses to the scalar partners of quarks and leptons. This approach has several advantages. The loop corrections can easily be large enough. They do not respect the mass relations that cause trouble at the tree level. The approach based on loop diagrams is most attractive in models in which the quarks, the leptons, and their scalar partners are all massless at the tree level. It is then quite natural for the loop corrections to scalar masses to be much bigger than the loop corrections to fermion masses. After all, the only approximate symmetry that the scalar masses violate is supersymmetry; but the fermion masses violate supersymmetry, SU(2) x U(1), and perhaps some approximate global symmetries. Moreover, models in which the quarks and leptons are massless at the tree level have another potential advantage. They could perhaps explain eventually why the quarks and leptons are so light compared to the W boson. Recently there have been developed some models in which the quark and lepton masses all come from loop diagrams[15], and other models in which loop diagrams are used to obtain realistic mass spectra[16].

Acknowledgement

I would like to thank the organizers of the UNAM school on supersymmetry for their hospitality.

References

1. H. Georgi, S. Glashow, Phys. Rev. Lett. 32 (1974) 438.
2. J.C. Pati, A. Salam, Phys. Rev D8 (1973) 1240.
3. M.S. Chanowitz, J. Ellis, M.K. Gaillard, Nucl. Phys. B128 (1977) 506; A.J. Buras, J. Ellis, M.K. Gaillard, D.V. Nanopoulos, Nucl. Phys. B135 (1978) 66.
4. H. Georgi, H. Quinn, and S. Weinber Phys. Rev. Lett. 33 (1974) 451. For a recent review, see W. Marciano and A. Sirlin, Proceedings of the Second Workshop on Grand Unification, Ann Arbor, April, 1981.
5. S. Weinberg, in Proc. of the XVII Int. Conf. on High Energy Physics, London, 1974, ed. J.R. Smith (Rutherford Laboratory, Berkshire, England, 1974).
6. P.A.M. Dirac, Proc. Roy. Soc. (London) A165 (1938) 199.
7. E. Witten, Nucl. Phys. B185 (1981) 513.
8. E. Witten, Phys. Lett. 105B (1981) 267; 1981 Erice Lecture notes, to appear.
9. E. Witten, "Constraints on Supersymmetry Breaking", Princeton preprint, submitted to Nuclear Physics B.
10. S. Dimopoulos, 1981 Erice lectures
11. W. Marciano, G. Senjanovic, preprint, Brookhaven National Laboratory; J. Ellis, D.V. Nanopoulus, Serge Rudaz, CERN preprint.
12. S. Dimopoulos, S. Raby, and F. Wilczek, Phys. Rev. D24 (1981) 1681.
13. S. Weinberg, Harvard preprint, 1981.
14. P. Fayet, Nucl. Phys. B90 (1975) 104, Phys. Lett. 58B (1975) 67; G.R. Farrar and P. Fayet, Phys. Lett. 76B (1978) 575, 79B (1978) 442, 89B (1980) 191.
15. C. Nappi, B. Ovrut, preprint, Institute for Advanced Study, to appear
16. M. Dine, W. Fischler, preprint, Institute for Advanced Study; L. Alvarez-Gaume, M. Claudson, and M. Wise, Harvard preprint.

YANG-MILLS THEORIES WITH GLOBAL AND LOCAL SUPERSYMMETRY - HIGGS AND SUPERHIGGS EFFECT IN UNIFIED FIELD THEORIES

S. Ferrara

Theory Division
CERN, 1211 Geneva 23
Switzerland

1. Introduction

The phenomenological success of the Glashow-Weinberg-Salam[1] model in describing electromagnetic and weak interactions in a unified fashion and in embedding the Fermi theory of weak interactions in a renormalizable field theory has dramatically reproposed unified field theories as the correct theoretical framework for describing all elementary particle interactions.

Fermions and bosons, the building blocks of matter, are the essential ingredients of any unified field theory based on renormalizable gauge interactions. Matter fields are usually described by spin 1/2 fermion fields while fields related to the geometry of the gauge group, such as the gauge vector bosons (connections) and Higgs fields, are described by spin 1 and spin 0 bosons.

In gravity, where the gauge principle is enlarged to the space-time symmetry, an additional bosonic tensor field, the spin 2 graviton, is also needed. In relativistic quantum mechanics the connection between spin and statistics and the underlying geometry of a gauge Lie group lead unavoidably to a dichotomy between matter and geometrical fields. More specifically, in the context of a Lie algebra symmetry it is impossible to achieve a complete unification of particle interactions due to the different spin (and statistics) of the elementary quanta which describe the fundamental forces of Nature.

Supersymmetry[2] is a space-time symmetry which embeds bosons and fermions in irreducible multiplets. This symmetry, although not manifest at present energies, has some appealing properties when applied to local quantum field theories and it may achieve the ultimate goal of a complete unification of particle interactions including gravitation[3]. Supersymmetry is in fact at present the only known symmetry consistent with local quantum field theory which relates particles of different spin[4]. This property allows previous theorems on the impossibility of non-trivial mixing between space-time and internal symmetries[5] to be overcome. Most surprisingly, supersymmetric field theories are the least divergent quantum field theories so far constructed. Supersymmetries are usually labelled by the number of fermionic invariances, i.e., by the number of spinorial generators $Q_\alpha{}^i$. When i = 1,...,N we call N-extended supersymmetry the corresponding supersymmetry algebra. For reasons which will become transparent in the next section, renormalizable local field theories admit up to N = 4 supersymmetries.

In N = 1 supersymmetry two kinds of multiplets and two kinds of interactions are possible. Chiral multiplets containing a left-handed (or Majorana) spin 1/2 fermion and a complex scalar can interact through a super-Yukawa interaction (to be described later). Vector multiplets containing a spin 1 vector particle and a left-handed (or Majorana) fermion are associated with ordinary gauge interactions, such as QED on Yang-Mills interactions, and can interact among themselves and with the chiral multiplets if the latter transform non-trivially under the gauge group.

In N = 2 supersymmetry[6] two kinds of multiplets also exist, but only gauge interactions are possible. Chiral multiplets contain in this case a Dirac spinor and two complex scalars, while vector multiplets contain a vector particle, a Dirac spinor and a complex scalar.

In N = 3 and 4 supersymmetry[7] only one multiplet, the vector multiplet, exists with the only possible Yang-Mills gauge self-interaction. This multiplet describes a vector particle, four left-handed (or Majorana) spin 1/2 fermions and six real scalars (or three complex ones).

Concerning the ultra-violet properties of these theories, particularly remarkable is the N = 4 Yang-Mills theory where the $\beta(g)$ function is known to vanish up to three loops by explicit calculations, and arguments have been given that it may in fact vanish to all orders of perturbation theory[8]. The softening of quantum divergences in supersymmetric Yang-Mills theories with respect to the ordinary renormalizable theories has recently proposed[9] these theories as candidates for the solution of the so-called hierarchy problem[10] that one encounters in grand unified theories of electroweak and strong interactions and even in the standard model of electroweak interactions[11]. In Section 4 we will describe attempts to use supersymmetric theories in the framework of GUT's and of electroweak interactions.

Some phenomenological problems which arise in this approach will also be considered. This survey is organized as follows: In section 2 we will describe the supersymmetry algebra and its representations on one-particle states. In Section 3 we will derive, using superspace, the most general $N = 1$ supersymmetry renormalizable Lagrangian and we will describe its component form. In Section 4 spontaneous supersymmetry breaking and attempts to build supersymmetric models of electroweak and strong interactions will be described. In Section 5 we will describe supersymmetric Yang-Mills theories coupled to supergravity and the superhiggs mechanism without a cosmological constant. We will show, in this last section how the supergravity breaking induced by purely supergravity couplings can solve the problem of mass splitting in broken supersymmetric theories.

2. Supersymmetry Algebras and Particle Multiplets

Supersymmetry is a graded Lie algebra (GLA) whose multiplication rules contain both commutators and anticommutators. In general, for GLA's one has

$$[X_A, X_B\} = f^C{}_{AB}\, X_C \text{ with } [X_A, X_B\} = X_A X_B - (-)^{AB} X_B X_A \quad , \tag{2.1}$$

where $A,B = 0$ if X is an even generator and $A,B = 1$ if X is an odd generator. The Jacobi identities are $[[X_A,X_B\},X_C\} = 0$ and $f^C{}_{AB}$ are the structure constants of the graded Lie algebra. The even part A of the X generators is an ordinary N-dimensional Lie algebra while the odd part S is the so-called grading representation of A. The GLA with generators $X = (A,S)$ has dimension $n + N$ where N is the dimension of the representation of A under which the S transforms.

In the case of supersymmetry the Lie algebra part has the structure $T \otimes G$ where T is the space-time symmetry and G is an internal symmetry group. The odd part S consists of a set of N self-conjugate spin 1/2 generators $Q_\alpha{}^i$ $(i=1,\dots,N)$, $(C(Q_\alpha)^T = Q_\alpha)$ which provide the grading of A.

The most interesting possibility most used in particle physics is the grading of the Poincaré algebra[4 12]. In this case the most general grading is given by

$$\{Q_\alpha{}^i, Q_\beta{}^j\} = -1/2(\gamma^\mu C)_{\alpha\beta}\, \delta^{ij} P_\mu + Z^{ij} C_{\alpha\beta} + \tilde{Z}^{ij} (\gamma_5 C)_{\alpha\beta} \quad ,$$

$$[Q_\alpha{}^i, M_{\mu\nu}] = (\sigma_{\mu\nu})_\alpha{}^\beta\, Q_\beta{}^i \quad , \tag{2.2}$$

$$[Q_\alpha{}^i, P_\mu] = [Q_\alpha{}^i, Z^{\ell m}] = [Q_\alpha{}^i, \tilde{Z}^{\ell m}] = [Z^{\ell m}, \tilde{Z}^{hk}] = 0 \quad .$$

The antisymmetric Poincaré invariant operators $Z^{ij}, \tilde{Z}^{ij}$ belong to the centre of the GLA. Obviously, they can only appear for N≥2. It is often convenient to rewrite (2.2) in terms of the chiral projections

$$(Q_\alpha{}^i)_L = 1/2(1 + \gamma_5)_\alpha{}^\beta\, Q_\beta{}^i \ , \ (Q_\alpha{}^i)_R = 1/2(1 - \gamma_5)_\alpha{}^\beta\, Q_\beta{}^i \quad . \tag{2.3}$$

In terms of chiral spinors Eq. (2.2) becomes

$$\begin{aligned}
\{(Q_\alpha{}^i)_L, (Q_\beta{}^j)_L\} &= 1/2[(1 + \gamma_5)C]_{\alpha\beta}\, (Z^{ij} + \tilde{Z}^{ij}) \quad , \\
\{(Q_\alpha{}^i)_R, (Q_\beta{}^j)_R\} &= 1/2[(1 - \gamma_5)C]_{\alpha\beta}\, (Z^{ij} - \tilde{Z}^{ij}) \quad , \\
\{(Q_\alpha{}^i)_L, (Q_\beta{}^j)_R\} &= -1/2[(1 + \gamma_5)\gamma^\mu C]_{\alpha\beta}\, P_\mu\, \delta^{ij} \quad .
\end{aligned} \tag{2.4}$$

It is also useful to use Van der Vaerden notation for chiral (Weyl) spinors[13]. One writes a four-component Majorana spinor as $Q = (Q_\alpha, \overline{Q}^{\dot\alpha})$ where $\overline{Q}^{\dot\alpha} = \varepsilon^{\dot\alpha\dot\beta}(Q_\beta)^*$ and $\varepsilon^{\alpha\beta} = \varepsilon^{\dot\alpha\dot\beta} = -\varepsilon_{\alpha\beta} = -\varepsilon_{\dot\alpha\dot\beta} = -\varepsilon^{\beta\alpha} = -\varepsilon^{\dot\beta\dot\alpha}$ $(\alpha, \beta = 1,2\ \ \dot\alpha, \dot\beta = 1,2)$. Then Eqs. (2.2) simplify as follows

$$\begin{aligned}
\{Q_\alpha{}^i, \overline{Q}_{\dot\beta}{}^j\} &= 1/2(\sigma^\mu)_{\alpha\dot\beta}\, P_\mu \quad , \\
\{Q_\alpha{}^i, Q_\beta{}^j\} &= i\varepsilon_{\alpha\beta}\, (Z^{ij} + \tilde{Z}^{ij}) \quad .
\end{aligned} \tag{2.5}$$

Once the algebra (2.4) (or (2.5)) is given, one can study its unitary irreducible representations using an extension of the Wigner method of induced representations[14]. In particular, one can study the general structure of the unitary (infinite dimensional) irreducible representations of the extended supersymmetry algebras acting on one particle states. We will consider first the algebra (2.4) in the zero central charge sector ($Z^{ij} = \tilde{Z}^{ij} = 0$).

For massive representations we can choose a Lorentz frame such that $P^\mu = (M,0)$ where M is the common mass of the multiplet components. Then (2.2) reduces to

$$\{Q_\alpha{}^i, Q_\beta{}^j\} = \delta_{\alpha\beta}\, \delta^{ij} \qquad \alpha, \beta = 1,\ldots,4 \ ; \ i,j = 1,\ldots,N \quad . \tag{2.6}$$

The even part is given by the angular momentum operator

$$J_k = \varepsilon_{k\ell m}\, M^{\ell m} \quad (\ell, m = 1,2,3) \quad . \tag{2.7}$$

The anticommutators (2.6) define the Clifford algebra for the group SO(4N). Its unique irreducible representation has dimension

2^{2N}. This is the spinor representation of SO(4N). Under SO(4N) this representation splits into two irreducible representations of dimension 2^{2N-1}. We can use two-component Weyl spinors as given in (2.5) and rewrite (2.6) as follows:

$$\{Q_\alpha{}^i, \overline{Q}_{\dot\beta}{}^j\} = \delta_{\alpha\dot\beta}\, \delta^{ij} \quad ,$$

$$\{Q_\alpha{}^i, Q_\beta{}^j\} = \{\overline{Q}_{\dot\alpha}{}^i, \overline{Q}_{\dot\beta}{}^j\} = 0 \quad . \tag{2.8}$$

The $Q_\alpha{}^i$ and $\overline{Q}_{\dot\alpha}{}^i$ operators satisfy the algebra of 2N fermionic creation and destruction operators. If we start from a state $|\Omega\rangle$ such that $Q^i|\Omega\rangle = 0$ (Clifford vacuum) then the previously alluded to 2^{2N} states are obtained as follows:

$$|\Omega\rangle\ ,\ \overline{Q}_{\dot\alpha}{}^i|\Omega\rangle\ ,\ \overline{Q}_{\dot\alpha_1}{}^{i_1}\ \overline{Q}_{\dot\alpha_2}{}^{i_2}\ |\Omega\rangle, \ldots,\ \overline{Q}_{\dot\alpha_1}{}^{i_1} \ldots,\ \overline{Q}_{\dot\alpha_k}{}^{i_k}\ |\Omega\rangle \ldots\ . \tag{2.9}$$

If we further define the 2N-component spinors

$$Q_\alpha{}^a = Q_\alpha{}^i \ , \quad (a = 1,\ldots,N,)$$

$$Q_\alpha{}^a = \overline{Q}^{\dot\alpha i} = \varepsilon^{\dot\alpha\dot\beta}\, \overline{Q}_{\dot\beta}{}^i \ , \quad (a = N+1,\ldots,2N) \quad , \tag{2.10}$$

(2.8) becomes

$$\{Q_\alpha{}^a, Q_\beta{}^b\} = \varepsilon_{\alpha\beta}\, \Omega^{ab} \ , \quad \alpha,\beta = 1,2 \quad , \quad a,b = 1,\ldots,2N \quad , \tag{2.11}$$

where

$$\Omega^{ab} = -\Omega^{ba} = \begin{bmatrix} 0 & I \\ -I & 0 \end{bmatrix} \ , \ (Q_\alpha{}^a)^* = \Omega_{ab}\, \varepsilon^{\alpha\beta}\, Q_\beta{}^b \quad .$$

The algebra (2.11) has manifest covariance under SU(2) x USp(2N) $\subset$ SO(4N). The 2^{2N} states of the spinor representation of SO(4N) decompose under SU(2) x USp(2N) as follows:

$$2^{2N} = (N+1,1) + (N,2N) + \ldots(N-k+1,[2N\text{x}\ldots 2N]_k) + \ldots$$
$$\ldots (1,[2N\ \text{x}\ldots 2N]_N) \quad . \tag{2.12}$$

The two irreducible representations of SO(4N) of dimension 2^{2N-1} correspond to integral and half-integral SU(2) spin, respectively. Then we have proven that the lowest dimensional massive irreducible representation of N extended supersymmetry contains spin states from $J = 0$ up to $J = N/2$. Higher dimensional irreducible representations are obtained by relaxing the condition that $|\Omega\rangle$ is an over-all singlet. In particular, $|\Omega\rangle$ can carry any value of the spin and transform in some non-trivial (real) representation of a subgroup $G \leq U(N)$ where $U(N)$ is the maximal group of automorphisms of the N extended supersymmetry algebra. Massive representations can be smaller in the presence of central charges. We give an example: in $N = 2n$ extended supersymmetry, massive (lowest dimensional) multiplets with a non-vanishing central charge have real dimension 2^{2N+1} instead of 2^{4N}. The spin runs from $J = 0$ up to $J = n/2$ (instead of $J = n$). These 2^{2N+1} states consist of a doublet of massive representations of n extended supersymmetry without central charges. The maximal group of automorphisms of the algebra is USp(2n) instead of U(2n).

In the application to quantum field theories the only interesting elementary massive multiplets are the lowest dimensional multiplet for $N = 1$ ($J = 0,1/2$) and the lowest dimensional multiplets with central charge for $N = 2$. According to the previous discussion this last multiplet consists of a pair of $J = 0,1/2$ chiral multiplets of $N = 1$ supersymmetry.

We now consider the more interesting case of massless representations. In this case we can choose a Lorentz frame $P^{\mu} = (P^o,0,0,P^o)$ for the light-like momentum P^{μ} ($P^{\mu}P_{\mu} = 0$, $P^o > 0$). The stability subalgebra, writen in terms of Weyl spinors, becomes

$$\{Q_{\alpha}{}^{i},\overline{Q}_{\dot\beta}{}^{j}\} = \left[\frac{1+\sigma_3}{2}\right]_{\alpha\dot\beta} \delta^{ij} \quad , \quad \{Q_{\alpha}{}^{i},Q_{\beta}{}^{j}\} = 0 \quad . \tag{2.13}$$

Equation (2.13) implies $Q_2{}^{i} = \overline{Q}_2{}^{i} = 0$. If we set $Q_1{}^{i} = Q^{i}$ we see that (2.13) becomes the Clifford algebra for SO(2N). Its unique irreducible representation has dimension 2^N which again splits into two irreducible representations of dimension 2^{N-1} under SO(2N). We are interested in this case in the decomposition of SO(2N) under the subgroup $U(1)_{helicity} \times SU(N)$. We have

$$2^N = (\lambda,1) + (\lambda - 1/2,\overline{N}) + \ldots(\lambda - k/2, [\overline{N}]_k) + \ldots(\lambda - N/2,1) \quad , \tag{2.14}$$

where λ is the helicity of the Clifford vacuum defined by $Q^i|\Omega\rangle = 0$. It is obvious that this irreducible representation does not have PCT-conjugate states because for the state of helicity λ there is no corresponding (antiparticle) state with helicity $-\lambda$.

This can only happen if $\lambda - N/2 = -\lambda$, i.e., $\lambda = N/4$. If $\lambda \neq N/4$ we must double the states and add to the multiplet with Clifford vacuum of helicity λ a PCT-conjugate multiplet with Clifford vacuum of helicity $N/2 - \lambda$. If the Clifford vacuum transforms under a non-trivial representation R of a subgroup $G \subset SU(N)$ then the PCT-conjugate Clifford vacuum must transform according to the complex conjugate representation $\overline{R}$. From the previous construction it follows that the minimum helicity range for a massless multiplet of N extended supersymmetry is from $\lambda = 0$ up to $\lambda = N/4$ ($(N+1)/4$ for odd N). Then if we confine ourselves to renormalizable interactions we find that $N \leq 4$ as claimed in the introduction. In the case of supergravity, when we allow for the spin 2 graviton to be the maximal helicity states, we have the bound $N \leq 8$.

3. Superspace, Multiplets of Fields and Interactions

In the application of supersymmetry to weak and electromagnetic interactions it is important to establish the internal symmetry properties of left-handed fermions.

If we confine ourselves to renormalizable Yang-Mills interactions we have seen that only three supersymmetry algebras with N = 1,2, and 4 are possible (the $\lambda_{MAX} = 1$ representation of N = 3 and N = 4 actually coincide). Furthermore, if we want left-handed fermions in complex representations of the gauge group G which commutes with the supersymmetry generators, then only one possibility is left, namely the N = 1 supersymmetry algebra. This is because for N = 4 fermions are partners of vector bosons and must therefore belong to the adjoint representation of G. For N = 2 the left-handed fermion belonging to the chiral multiplet has a right-handed partner, both belonging to the same (arbitrary) representation R of G.

For N = 1, the chiral multiplet describes just one left-hand spin 1/2 fermion which can therefore belong to a complex ($R \neq \overline{R}$) representation of the gauge group G.

Let us consider the relevant field representations of N = 1 supersymmetry. The simplest one is given by the chiral multiplet[5]

$$S = (A(x),\ B(x),\ \chi(x),\ F(x),\ G(x)) \quad . \tag{3.1}$$

A, B, F, G are four real scalar fields and $\chi(x)$ is a Majorana spinor. The infinitesimal transformation laws under supersymmetry are

$$\delta A = i\bar{\varepsilon}\chi \ , \ \delta B = -\bar{\varepsilon}\gamma_5\chi \ , \ \delta\chi = \partial\!\!\!/(A + i\gamma_5 B)\varepsilon + (F + i\gamma_5 G)\varepsilon,$$

$$\delta F = i\bar{\varepsilon}\partial\!\!\!/\chi \ , \ \delta G = -\bar{\varepsilon}\gamma_5\partial\!\!\!/\chi \quad . \tag{3.2}$$

ε_α is a constant anticommuting Majorana spinor. It is useful to introduce complex notation:

$$Z = \frac{A-iB}{2} \quad , \; \chi = 1/2(1+\gamma_5)\chi \; , \quad H = \frac{F+iG}{2} \quad . \tag{3.3}$$

Then (3.2) becomes

$$\begin{aligned} \delta Z &= \varepsilon^\alpha \chi_\alpha \quad , \\ \delta \chi_\alpha &= 2i(\sigma^\mu)_{\alpha\dot\beta}\, \bar\varepsilon^{\dot\beta}\, \partial_\mu Z + 2H\, \varepsilon_\alpha \quad , \\ \delta H &= -i\partial_\mu \chi^\alpha\, (\sigma^\mu)_{\alpha\dot\beta}\, \bar\varepsilon^{\dot\beta} \quad . \end{aligned} \tag{3.4}$$

To include vector particles a larger multiplet must be introduced[15] (vector multiplets). Its real components are

$$V = (C,\zeta,M,N,v_\mu,\lambda,D) \quad . \tag{3.5}$$

C,M,N,D are real scalars, ζ and λ are Majorana spinors and v_μ a real vector field. The infinitesimal supersymmetry transformations are

$$\begin{aligned} \delta C &= -\bar\varepsilon\gamma_5\zeta \quad , \\ \delta\zeta &= \gamma^\mu v_\mu \varepsilon - i\partial_\mu C\, \gamma_5\gamma^\mu\, \varepsilon + (M + i\gamma_5 N)\varepsilon \quad , \\ \delta M &= -i\bar\varepsilon\lambda + i\bar\varepsilon\not\partial\zeta \quad , \\ \delta N &= -\bar\varepsilon\gamma_5\lambda - \bar\varepsilon\gamma_5\not\partial\zeta \quad , \\ \delta v_\mu &= i\bar\varepsilon\gamma_\mu\lambda + i\bar\varepsilon\partial_\mu\zeta \quad , \\ \delta\lambda &= -1/2\; F_{\mu\nu}\, \gamma^\mu\gamma^\nu\, \varepsilon + iD\gamma_5\varepsilon \quad , \\ \delta D &= -\bar\varepsilon\gamma_5\not\partial\lambda \quad . \end{aligned} \tag{3.6}$$

This multiplet can be subjected to a supersymmetric U(1) gauge transformation by observing that the following multiplet[16]

$$C = B \; , \; \zeta = \chi \; , \; M = F \; , \; N = G \; , \; v_\mu = \partial_\mu A \; , \; \lambda = D = 0 \quad , \tag{3.7}$$

transforms as in (3.6). Then we can consistently demand that

$$\begin{aligned} \delta C &= B \; , & \delta v_\mu &= \partial_\mu A \; , \\ \delta\zeta &= \chi \; , & \delta\lambda &= 0 \; , \\ \delta M &= F \; , & \delta D &= 0 \; , \end{aligned}$$

$$\delta N = G \ . \tag{3.8}$$

Since the transformation (3.8) acts as a translation on the first four components V we can set them equal to zero by fixing a supersymmetric gauge and we are left with

$$\delta v_\mu = \partial_\mu A \ , \ \delta\lambda = \delta D = 0 \quad , \tag{3.9}$$

i.e., an ordinary gauge transformation on v_μ, while λ and D are gauge invariant. The gauge for which $C = \zeta = M = N = 0$ is called the Wess-Zumino gauge. In this gauge we have

$$\delta v_\mu = i\bar{\varepsilon}\gamma_\mu\lambda \ , \quad \delta\lambda = -F_{\mu\nu}\,\sigma^{\mu\nu}\varepsilon \ , \ \delta D = -\bar{\varepsilon}\gamma_5\not{\partial}\lambda \quad . \tag{3.10}$$

The supersymmetry algebra is modified by the inclusion of gauge transformations, i.e.,

$$[\delta_1,\delta_2]v_\mu = -2i\ \bar{\varepsilon}_1\gamma^\nu\varepsilon_2\ F_{\nu\mu} \quad . \tag{3.11}$$

For non-Abelian gauge transformations in the Wess-Zumino gauge, we have

$$\begin{aligned} \delta v_\mu &= i\bar{\varepsilon}\gamma_\mu\lambda \quad , \\ \delta\lambda &= -G_{\mu\nu}\,\sigma^{\mu\nu}\varepsilon + i\gamma_5 D\varepsilon \quad , \\ \delta D &= -\bar{\varepsilon}\gamma_5\not{D}\lambda \quad , \end{aligned} \tag{3.12}$$

where $G_{\mu\nu}$ is the non-Abelian field strength of v_μ, D_μ is the covariant derivative and λ and D belong to the adjoint representation of the gauge group G.

Let us now suppose that the chiral multiplet S belongs to some irreducible representation R of G. Its supersymmetry transformation in the Wess-Zumino gauge becomes:

$$\begin{aligned} \delta Z^a &= \varepsilon^\alpha\ \chi_\alpha{}^a \quad , \\ \delta\chi_\alpha{}^a &= 2i(\sigma^\mu)_{\alpha\dot\beta}\ \bar{\varepsilon}^{\dot\beta}(D_\mu Z)^a + 2H^a\ \varepsilon_\alpha \quad , \\ \delta H^a &= -i(D_\mu\chi^\alpha)^a\ (\sigma^\mu)_{\alpha\dot\beta}\ \bar{\varepsilon}^{\dot\beta} - 2ig\ \bar{\lambda}_{\dot\alpha}{}^A\ \bar{\varepsilon}^{\dot\alpha}(R^A Z)^a \quad . \end{aligned} \tag{3.13}$$

In order to establish interactions among the two types of multiplets we have introduced, it is convenient to go to superspace[17] where the supersymmetric tensor calculus becomes transparent.

Superspace is an extension of ordinary space-time to a space-time with spin degrees of freedom. The basic manifold of superspace for N = 1 supersymmetry has points parametrized by coordinates

$$Z^A = (x^\mu, \theta^\alpha, \overline{\theta}^{\dot\alpha}) \qquad \mu = 1,\ldots,4 \qquad \alpha,\dot\alpha = 1,2 \quad . \tag{3.14}$$

The θ variables have to be considered as odd elements of a Grassmann algebra, i.e., $[Z^A, Z^B\} = 0$.

Superspace is the quotient space G/H where G is the 14-dimensional graded Poincaré algebra (N = 1 supersymmetry algebra) and H is the homogeneous Lorentz group.

Supersymmetry transformations are realized as motions in superspace. Under an infinitesimal supertranslation of parameters $(a^\mu, \varepsilon^\alpha, \overline{\varepsilon}^{\dot\alpha})$ we have

$$\begin{aligned} \delta x^\mu &= -i\varepsilon\sigma^\mu\overline{\theta} + i\theta\sigma^\mu\overline{\varepsilon} + a^\mu \quad , \\ \delta\theta^\alpha &= \varepsilon^\alpha \ , \ \delta\overline{\theta}^{\dot\alpha} = \overline{\varepsilon}^{\dot\alpha} \quad . \end{aligned} \tag{3.15}$$

The composition rule of two supertranslations of parameters $(a_1{}^\mu, \varepsilon_1{}^\alpha, \overline{\varepsilon}_1{}^{\dot\alpha}, a_2{}^\mu, \varepsilon_2{}^\alpha, \overline{\varepsilon}_2{}^{\dot\alpha})$ gives

$$[\delta_2, \delta_1] Z^A = (-2i\varepsilon_1\sigma^\mu\overline{\varepsilon}_2 + 2i\varepsilon_2\sigma^\mu\overline{\varepsilon}_1, 0, 0) \quad . \tag{3.16}$$

Superspace provides a representation of the supersymmetry algebra in terms of differential operators. The motion (3.15) is obtained in a standard way as the left action of a group element on the coset space G/H. The infinitesimal generators of this motion are

$$\begin{aligned} P_\mu &= i\frac{\partial}{\partial x^\mu} \quad , \quad Q_\alpha = \frac{\partial}{\partial\theta^\alpha} - i(\sigma^\mu)_{\alpha\dot\beta}\,\overline{\theta}^{\dot\beta}\frac{\partial}{\partial x^\mu} \quad , \\ \overline{Q}_{\dot\alpha} &= -\frac{\partial}{\partial\overline{\theta}^{\dot\alpha}} + i\theta^\beta(\sigma^\mu)_{\beta\dot\alpha}\frac{\partial}{\partial x^\mu} \end{aligned} \tag{3.17}$$

A scalar superfield is a scalar function in superspace, i.e., $\phi'(x'\theta') = \phi(x,\theta)$. In the infinitesimal one has

$$\delta\phi = \left[\varepsilon\frac{\partial}{\partial\theta} + \overline{\varepsilon}\frac{\partial}{\partial\overline{\theta}} + i(\theta\sigma^\mu\overline{\varepsilon} - \varepsilon\sigma^\mu\overline{\theta})\partial_\mu\right]\phi \quad . \tag{3.18}$$

For a superfield operator, $\delta\phi$ stands for $[\varepsilon^\alpha Q_\alpha + \overline{\varepsilon}_{\dot\alpha}\overline{Q}^{\dot\alpha}, \phi]$; $\phi(x,\theta)$ is a finite collection of ordinary local fields because $\theta_\alpha\theta_\beta = -\theta_\beta\theta_\alpha$ hence $\theta_{\alpha_1}\cdots\theta_{\alpha_n} = 0$ for $n > 4$. It follows that

$$\phi(x,\theta) = \sum_{i=1}^{4} \theta^{\alpha_1}\cdots\theta^{\alpha_i}\ \phi_{\alpha_1\ldots\alpha_i}(x) \quad . \tag{3.19}$$

From (3.18) one can compute $\delta\phi_{\alpha_1\ldots\alpha_i}(x)$ and get symbolically

$$\delta_\varepsilon \phi_{\alpha_1 \ldots \alpha_i}(x) \to \phi_{\alpha_1 \ldots \alpha_{i+1}}(x) + \partial_x \phi_{\alpha_1 \ldots \alpha_{i-1}}(x) \quad . \tag{3.20}$$

Because of (3.16) it is evident that superspace has non-vanishing (super) torsion. However, one can construct covariant differential operators[13] which commute with supersymmetry variations. They are

$$D_\alpha = \frac{\partial}{\partial \theta^\alpha} + i(\sigma^\mu)_{\alpha\dot\beta}\, \overline{\theta}^{\dot\beta} \frac{\partial}{\partial x^\mu} ,$$

$$\overline{D}_{\dot\alpha} = -\frac{\partial}{\partial \overline{\theta}^{\dot\alpha}} - i\theta^\beta (\sigma^\mu)_{\beta\dot\alpha} \frac{\partial}{\partial x^\mu} \quad . \tag{3.21}$$

Comparing (3.17) with (3.21) it is obvious that

$$D_\alpha \delta\phi = \delta D_\alpha \phi \quad . \tag{3.22}$$

The covariant derivatives satisfy the algebra

$$\{D_\alpha, \overline{D}_{\dot\alpha}\} = -2i(\sigma^\mu)_{\alpha\dot\alpha}\, \partial_\mu , \quad \{D_\alpha D_\beta\} = \{\overline{D}_{\dot\alpha}, \overline{D}_{\dot\beta}\} = 0 \quad ,$$

$$[D_\alpha, \partial_\mu] = [\overline{D}_{\dot\alpha}, \partial_\mu] = 0 \quad . \tag{3.23}$$

From (3.23) it also follows that $D_\alpha D_\beta D_\gamma = 0$. We can impose on a complex superfield the condition

$$\overline{D}_{\dot\alpha}\, S(x,\theta,\overline{\theta}) = 0 \quad , \tag{3.24}$$

which is solved by

$$S(x - i\sigma\overline{\theta}, \theta, \overline{\theta}) = \psi(x,\theta) = Z(x) + \theta^\alpha \chi_\alpha + \theta^\alpha \theta_\alpha\, H(x) \quad . \tag{3.25}$$

Equation (3.25) gives the chiral superfield S as defined in (3.4). In order to obtain the vector multiplet V it is sufficient to take a general scalar superfield subjected to the reality condition $\phi = \phi^*$. This condition is invariant under supersymmetry because δ is a real transformation, i.e., $(\delta\phi)^* = \delta(\phi^*)$. The previous analysis can be extended to an arbitrary superfield endowed with Lorentz and internal symmetry indices. A general superfield operator can be written as

$$\phi(x,\theta) = L(x,\theta)\ \phi\ L^{-1}(x,\theta) \quad , \tag{3.26}$$

where $L(x,\theta)$ is an element of the 10-dimensional coset space G/H. By definition $\phi = \phi(0,0)$. ϕ is a representation H. Note that H leaves the superorigin $x = \theta = 0$ invariant. If ϕ is a representation of H we can induce on $\phi(x,\theta)$ a representation of G. If X is a generator of H the action of X on $\phi(x,\theta)$ is computed as follows

$$[X, e^{-T}\,\phi\, e^{T}] = e^{-T}[Y,\phi]e^{T} + e^{-T}[X,\phi]e^{T} \quad , \tag{3.27}$$

where

$$Y = \int_0^1 d\lambda\; e^{\lambda T}[T,X]e^{-\lambda T} \;,\quad T = i(\theta^{\alpha}Q_{\alpha} + \bar{\theta}_{\dot\alpha}\bar{Q}^{\dot\alpha} - x^{\mu}P_{\mu}) \quad .$$

The infinite chain of commutators which defines Y stops after a finite number of steps because of the O'Raifeartaigh theorem.

We now attack the problem of constructing invariant interactions in a superfield formalism. A scalar field, in order to be a candidate for a supersymmetric action, must transform as a total divergence under superymmetry variations. Because of (3.20) this is achieved by the last component of a superfield. If we use the integration over anticommuting variables, defined by the Berezin recipe

$$\int d\theta_i \theta_j = \delta_{ij} \quad , \tag{3.28}$$

we can rewrite the last component of a superfield as

$$\int d^4\theta\; \phi(x,\theta) \quad \text{or} \quad \int d^2\theta\; \phi(x,\theta) \quad \text{if } \bar{D}_{\dot\alpha}\phi = 0 \quad . \tag{3.29}$$

Given a chiral superfield S ($\bar{D}_{\dot\alpha}S = 0$) we can write two possible invariant bilinears as follows

$$\int d^4x \int d^4\theta\; S\bar{S} \quad , \tag{3.30}$$

$$\int d^4x \int d^2\theta\; S^2 \quad . \tag{3.31}$$

Because of the property $\int d\theta = \partial/\partial\theta$, (3.30) may also be rewritten as

$$\int d^4x \int d^2\theta\; S\bar{D}\bar{D}\bar{S} \quad , \tag{3.32}$$

(the imaginary part of SDDS being a total divergence).

Because dim S = 1 (in energy units) we see that dim $S\bar{S}$ = 2 and dim $\int d^4\theta\; S\bar{S}$ = 4 (dim $d\theta$ = dim $\partial/\partial\theta$ = 1/2). Analogously dim $\int d^2\theta\; S^2$ = 3.

The only interacting term consistent with renormalizability is

$$\int d^2\theta\; S^3 \quad , \tag{3.33}$$

which has dimension four. An additional term is also possible with dimension two, namely

$$\int d^2\theta\; S \quad . \tag{3.34}$$

Therefore we conclude that the most general Lagrangian of interacting chiral multiplets consistent with renormalizability (dim L $\leq$ 4) is given by

$$\int d^4\theta \; S^a \bar{S}^a + \mathrm{Re} \int d^2\theta \; f(S^a) \quad , \qquad (3.35)$$

where

$$f(S^a) = \eta^a S_a + m^{ab} \, S_a S_b + g^{abc} \, S_a S_b S_c \quad .$$

$f(S_a)$ is often called the superpotential function. If a group G is acting on the index "a" one then gets a further constraint on f from the requirement of G invariance, namely

$$f_{,a}(S) \; (T_\alpha)^a{}_b \; S^b = 0 \quad , \qquad (3.36)$$

where $(T_\alpha)^a{}_b$ are the generators of the representation R of G under which S^a transforms.

In component form we have for the two terms in (3.33) respectively

$$L_{KIN} = -2\partial_\mu Z^a \; \partial^\mu Z_a{}^* - i\chi^{\alpha a} \; (\sigma_\mu)_{\alpha\dot{\beta}} \; \partial_\mu \bar{\chi}^{\dot{\beta}}{}_a + 2H^a H_a \quad ,$$

$$(3.37)$$

$$L_{pot} = H^a \; f_{,a} - 1/4 \; \chi^{\alpha a} \; \chi_\alpha{}^b \; f_{,ab} + \text{h.c.} \quad ,$$

when Z, χ and H have been defined in (3.3).

Elimination of H^a yields to the final component Lagrangian

$$L = -2\partial_\mu Z^a \; \partial^\mu Z_a{}^* - i\chi^{\alpha a} \; (\sigma_\mu)_{\alpha\dot{\beta}} \; \partial_\mu \bar{\chi}^{\dot{\beta}}{}_a$$
$$- 1/2 f_{,a} \; f^{*,a} - 1/4\chi^{\alpha a} \; \chi_\alpha{}^b \; f_{,ab} - 1/4 \; \bar{\chi}_{\dot{\alpha} a} \; \bar{\chi}^{\dot{\alpha}}{}_b \; f^{*,ab} \quad .$$

$$(3.38)$$

The scalar potential is given by

$$V = 1/2 \; \left| f_{,a} \right|^2 \qquad (3.39)$$

and it is semi-positive definite.

Let us now consider gauge theories[18], with a vector multiplet V. Let us endow V with an index A when A = 1,...,D and D is the dimension of the Lie algebra of a compact Lie group G. (We take SU(N) as an illustrative example). More specifically we can assume that V belongs to the Lie algebra G, i.e., is a Lie algebra valued function. If we introduce a chiral superfield Λ ($\bar{D}_{\dot{\alpha}}\Lambda = 0$), which is also Lie algebra valued, then a finite Yang-Mills transformation in superspace becomes

$$e^{2gV} \to e^{-i\Lambda^\dagger} e^{2gV} e^{i\Lambda} \tag{3.40}$$

where V, Λ and $\Lambda^\dagger$ are N x N Hermitean traceless matrices. It is easy to see that the chiral superfield

$$W_\alpha = 1/g^2\ \overline{D}_{\dot\alpha}\overline{D}^{\dot\alpha}(e^{-2gV}\ D_\alpha\ e^{2gV}) \tag{3.41}$$

transforms as

$$W_\alpha \to e^{-i\Lambda}\ W_\alpha\ e^{i\Lambda} \tag{3.42}$$

under a Yang-Mills transformation. W_α is the super-Yang-Mills strength in the sense that it contains only the Yang-Mills covariant quantities λ_α, $(\sigma^\mu)_{\alpha\dot\alpha}\ \tilde{D}_\mu\ \overline{\lambda}^{\dot\alpha}$ and $G_{\mu\nu} = \partial_\mu v_\nu - \partial_\nu v_\mu + ig\,[v_\mu, v_\nu]$. Here $\tilde{D}_\mu$ is the Yang-Mills covariant derivative $\tilde{D}^\mu\lambda = \partial_\mu\lambda + ig[v_\mu, v_\nu]$.

If V is a singlet superfield (3.40) and (3.42) simplify considerably and we obtain an Abelian U(1) gauge transformation and gauge field strenght respectively

$$V \to V + (1/2g)i(\Lambda - \overline{\Lambda}) \quad ,$$

$$W_\alpha = \overline{D}_{\dot\alpha}\ \overline{D}^{\dot\alpha}\ D_\alpha V \quad . \tag{3.43}$$

In terms of W_α the pure Yang-Mills supersymmetric action is given by

$$\int d^4x \int d^2\theta\ \mathrm{Tr}\ W^\alpha\ W_\alpha \quad . \tag{3.44}$$

Let us now consider a (matter) chiral superfield S^a ($\overline{D}_{\dot\alpha}\ S^a = 0$) belonging to some irreducible representations R of G. Then under a Yang-Mills transformation it transforms as

$$S \to e^{-i\Lambda}\ S \quad ,$$

$$\overline{S} \to \overline{S}\ e^{i\Lambda^\dagger} \quad , \tag{3.45}$$

when the Lie algebra valued parameter Λ now takes values in the representation space of S. One sees inmmediately that the following quantitiy

$$\overline{S}\ e^{2gV}\ S = \overline{S}_a (e^{2gV})^a{}_b\ S^b \tag{3.46}$$

is a Yang-Mills invariant and defines a vector multiplet whose last component gives rise to the supersymmetric minimal coupling. The most general superfield Lagrangian is therefore given by

$$\int d^4x[\int d^4\theta\ \overline{S}\ e^{2gV}S + \mathrm{Re} \int d^2\theta[\mathrm{Tr}(W^\alpha\ W_\alpha) + f(S)] \quad , \tag{3.47}$$

where S denotes a collection of irreducible representations R^i of G and f(S) is a G invariant superpotential function as defined by (3.35) and (3.36). If the gauge group G contains p Abelian factors $U(1)^p$, p additional terms are also possible, namely

$$\int d^4x \int d^4\theta \sum_{i=1}^{p} \xi^i V_i \quad , \tag{3.48}$$

where V_i are the vector superfields associated with the p U(1) factors of G.

In component form (3.44) becomes

$$\mathrm{Tr}\left(-1/4\ (G_{\mu\nu})^2 - i/2\ \overline{\lambda}\gamma^\mu \tilde{D}_\mu\lambda + 1/2\ D^2\right) \quad . \tag{3.49}$$

The component form of (3.46) gives

$$-2\left|\tilde{D}_\mu Z\right|^2 - i\chi^{\alpha a}(\sigma^\mu)_{\alpha\dot{\beta}}(\tilde{D}_\mu\overline{\chi}^{\dot{\beta}})_a + 2\left|H_a\right|^2$$

$$+2gi\ \lambda^{\alpha A}\ \chi_\alpha{}^a\ R^A{}_a{}^b\ Z_b{}^* + 2g\ Z^a\ R^A{}_a{}^b\ Z_b{}^*\ D^A + \text{h.c.} \tag{3.50}$$

while $R^A{}_a{}^b$ are the Hermitean generators of the representation R of S^a. If more irreducible representations are present in (3.50) a sum over all inequivalent irreducible representations is understood. $\tilde{D}_\mu$ is the ordinary Yang-Mills covariant derivative, i.e.,

$$\tilde{D}_\mu Z^a = \partial_\mu Z^a + ig\ v^A{}_\mu\ R^A{}_b{}^a\ Z^b \quad .$$

4. Spontaneous Supersymmetry Breaking and Application to Unified Theories

In the present section we describe some attempts to use N = 1 supersymmetric gauge theories to describe models of electroweak and strong interactions. To have a realistic model we have first to discuss spontaneous supersymmetry breaking. In a quantum field theory the condition for symmetry breaking for a given generator X of a continuous symmetry is that $X|0\rangle \neq 0$. In the case of supersymmetry this implies that $Q_\alpha|0\rangle \neq 0$. From the transformation laws (3.4) and (3.6) we see that this demands that H^a or (and) D^A must acquire a non-vanishing expectation value on the classical solutions of the field equations. In a gauge theory defined by the Lagrangian (3.47) (with the possible modification given by (3.48)) this requires that one of these equations

$$f_{,a} = 0 \tag{4.1}$$

$$D^A = -2g\ Z^a\ R^A{}_a{}^b\ Z_b - \xi_i\delta^{Ai} = 0 \quad (i = 1,\ldots p) \quad , \tag{4.2}$$

does not admit a solution.

It is evident that if all chiral multiplets transform under G ($\eta^a = 0$ in (3.35)) and if $\xi_i = 0$, Equations (4.1) have always the G invariant solution $Z^a = 0$. The only way to allow for an inconsistency of (4.1) and (or) (4.2) is to have some singlet fields X^a under G ($\eta^a \neq 0$) or (and) some U(1) factor in G with $\xi_i \neq 0$. Therefore, a necessary condition to have spontaneous supersymmetry breaking at the tree-level in a supersymmetric gauge theory is to have one of the following two conditions satisfied:

i) The group G should contain at least a neutral field X with linear term in the superpotential f(S).

ii) The group G should contain at least an Abelian factor U(1) with a non-vanishing ξ in (3.48).

One can easily see that i) and ii) are necessary but not sufficient conditions to have spontaneous supersymmetry breaking at the tree level.

Let us consider an example which fullfills condition i). Consider a group G, a singlet X and two chiral superfields S_1, S_2 transforming according to a real representation R of G.

Then the following superpotential term[19]

$$f(S_i,X) = g\,X\,(S_1)^2 + \mu^2 X + m\,S_1\,S_2$$

undergoes spontaneous supersymmetry breaking. In fact we get

$$\begin{aligned} f_{,X} &= g\,(S_1)^2 + \mu^2 \quad , \\ f_{,S_1} &= m\,S_2 + 2g\,X\,S_1 \quad , \\ f_{,S_2} &= m\,S_1 \end{aligned} \qquad (4.3)$$

and we see that the two equations $f_{,X} = 0$ and $f_{,S_2} = 0$ are incompatible. An obvious property of this model is that det $f_{,ab} = 0$ and that the tree level potential does not fix the three vacuum expectation values $\langle x\rangle$, $\langle S_1\rangle$ and $\langle S_2\rangle$. Let us consider the full gauge theory with the over-all potential

$$1/2\ f_{,a}\,f^*_{,a} + 1/2\ D^A\,D^A \qquad (4.4)$$

where

$$D^A = -2g\ Z^a\,R^A{}_a{}^b\,Z_b^* - \xi_i\delta^{Ai} \qquad (i = 1,\dots p)$$

Differentiating (4.4) we get the extremum condition

$$f_{,ab}\, f^{*a} + 2D^A\, D^A{}_{,b} = 0 \quad . \tag{4.5}$$

Equation (4.5) can be written as

$$M_{AB}\, \tilde{F}^B = 0 \quad , \tag{4.6}$$

where

$$\tilde{F}^B = (f^{*a}, D^A)$$

and M_{AB} is the matrix

$$M_{AB} = \begin{bmatrix} f_{,ab} & 2D^A{}_{,b} \\ 2D^A{}_{,a} & 0 \end{bmatrix} \quad .$$

The condition that (4.6) has a solution with a non-vanishing eigenvector $\tilde{F}^B$ implies[20] that det $M_{AB} = 0$, i.e., the fermion mass matrix must have a vanishing eigenvalue, the Goldstone fermion. The Goldstone fermion is a combination of $f_{,a}\,\chi^a$ and $D^A\,\lambda^A$ with strength fixed by (4.5) and (4.6). Therefore, the Goldstone fermion is the mixture of those left-handed fermions $\chi^a{}_L, \lambda^A$ for which the corresponding auxiliary fields $f_{,a}$ and D^A are different from zero at the absolute minimum of the potential determined by (4.5).

The main application of supersymmetric Yang-Mills theories to low energy physics is in the context of grand unified theories and of the conventional Weinberg-Salam model.

In GUT's one is faced with the hierarchy problem[10,11] which is connected with the fact that there is no natural way of having a small expectation value (or Higgs mass) for the Higgs weak SU(2) doublet due to quadratically divergent self-energy graphs which mix the SU(2) x U(1) breaking scale $M_W \sim 100$ GeV with the grand unification scale M_X (Planck scale M_{Pl}?). In supersymmetric gauge theories, because of non-renormalization theorems[21], it is known that it is "natural" to set a scalar mass equal to zero. If we stipulate that the supersymmetry breaking be related to the SU(2) x U(1) breaking of the electroweak interaction group then one can find a perhaps unique mechanism that could explain naturally the smallness of the Higgs boson mass in spite of the fact that the complete theory contains a big scale M_X [9] (or M_{Pl}). The only supersymmetric theories which have quadratic divergences[22] are theories where the gauge group G contains a $\tilde{U}(1)$ group with a non-vanishing ξV term as defined by (3.48) and with an over-all $\tilde{U}(1)$ non-vanishing trace over the chiral fields S^a

$$\mathrm{Tr}_a\, \tilde{Y} \neq 0 \quad . \tag{4.7}$$

It turns out that in order to have a realistic mass spectrum for quarks, leptons and their scalar superpartners the effective gauge theory must contain at least the group[23,24]

$$SU(3) \times SU(2) \times U(1) \times \tilde{U}(1) \quad , \tag{4.8}$$

with the value of the $\tilde{Y}$ charge which is positive (or negative) over all left-handed matter fields. This turns out to be a consequence of the general mass matrix for quarks and leptons as derived by (3.50) under the assumption that colour and electromagnetism are unbroken symmetries at low energies. If we want to avoid quadratic divergences then in addition we must have[20,22]

$$Tr_a \tilde{Y} = 0 \quad , \tag{4.9}$$

which means that other fields must exist, carrying colour and SU(2) quantum numbers with opposite values of the $\tilde{Y}$ charge with respect to the matter fields. The fact that these extra fields carry colour and charge does not come from (4.9) but from the triangle anomaly equations of the type

$$Tr\,(SU(2)^2\,\tilde{U}(1)) = Tr\,(SU(3)^2\,\tilde{U}(1)) = Tr\,(\tilde{U}(1)^3) = 0 \quad . \tag{4.10}$$

For each generation of quarks and leptons the first equation can be satisfied by the Higgs doublets, the second equation by a colour triplet and a colour antitriplet and the third equation by two additional scalars which are neutral under SU(3) x SU(2) x U(1). However, one can show that even neglecting matter the most general potential, involving the previously mentioned 27 fields which may be embedded in the 27-dimensional representation of E_6, is not able to stabilize the colour triplet fields then leading to an unacceptable situation where colour is broken.

Recently Weinberg has given some criteria for finding physically relevant local minima in a general supersymmetric gauge theory. Unfortunately these criteria do not exclude the existence of lower supersymmetric minima, for any range of the parameters of the theory, which have physically unwanted properties[25]. It is possible to show that by extending the 27-dimensional representation of E_6 whose reduction under SU(3) x SU(2) x U(1) x $\tilde{U}(1)$ is[25]

$$\begin{aligned} &Q(3,2,1/6,1)\ ;\ u^c(\bar{3},1,-2/3,1)\ ;\ d^c(\bar{3},1,1/3,1);\\ &L(1,2,-1/2,1)\ ;\ e^c(1,1,1,1)\ ;\ P(1,1,0,1);\\ &H(1,2,-1/2,-2)\ ;\ H^c(1,2,1/2,-2)\ ;\ T(3,1,-1/3,-2)\ ;\\ &\qquad T^c(\bar{3},1,1/3,-2)\ ;\ S(1,1,0,4) \quad , \end{aligned} \tag{4.11}$$

to a larger set of 30 fields by adding three chiral fields singlets under SU(3) x SU(2) x U(1) with $\tilde{U}(1)$ charge given by

$$R(1,1,0,4)\ ,\ R^c(1,1,0,-4)\ ,\ N(1,1,0,0)\ , \qquad (4.12)$$

it is possible to find a local minimum with all desired properties, which stabilizes the colour triplets and therefore does not break colour and charge[25].

Unfortunately, this model has always a supersymmetric solution which breaks colour and charge. It is also possible to show that any modification of this set of 30 chiral multiplets by addition of $\tilde{U}(1)$ anomaly free and traceless sets of chiral superfiels not transforming under SU(3) x SU(2) x U(1) does not help. A possible way out is to enlarge the gauge group to SU(3) x SU(2) x U(1) x $\tilde{U}(1)$ x U'(1) and to play with the three parameters ξ_Y, ξ and ξ' to see whether supersymmetric solutions can be avoided and physically acceptable supersymmetric breaking solutions exist.

A possible way out of these difficulties has recently been proposed[25] in a class of models in which the group $\tilde{U}(1)$ is broken at a very high scale ξ, while the SU(2) x U(1) group and supersymmetry are controlled by a much smaller scale μ. Higgs and boson squared masses are found of order μ^2 in these models while the U(1) vector boson squared mass is of order ξ and the gravitino squared mass is of order $\mu^2\xi/(M_{Pl})^2$. Since the $\tilde{U}(1)$ group is broken at a high scale ξ ($\sim M_{Pl}$) the effect of possible anomalies is not dramatic and in any case does not disturb the low energy theory controlled by the much smaller scale μ. Interestingly enough, if $\xi \sim (M_{Pl})^2$ the gravitino has a mass of order μ and it is almost decoupled from low energy physics.

To illustrate the model we just consider the relevant fields which are H, H^c, S, R, and R^c. The superpotential is

$$H\, H^c\, S + \mu\, R\, R^c \qquad (4.13)$$

and supersymmetry is broken if $\mu \neq 0$. The complete model is obtained with the 30 scalar fields previously described with the exclusion of T and T^c. Of course, the model has $\tilde{U}(1)$ anomalies. However, since $\tilde{U}(1)$ can be broken at super high energies where gravitational interactions become strong, the presence of these anomalies is not dramatic. Finally, it should be mentioned that even in a completely anomaly-free theory with a small scale ξ, gravitational interactions would introduce triangular anomalies and therefore not renormalizable effects due to a modification of the $\tilde{U}(1)$ charge which is proportional to $\xi/(M_{Pl})^2$.

As a final comment we should mention what the scenario would be if the group G were semi-simple. In this case the terms $\xi^k V_k$ in the superspace Lagrangian (see (3.48)) are absent and the only possible supersymmetry breaking at the tree level is obtained by a mechanism as given by i). It is a general property of such a kind of potentials that the mass spectrum of the scalar partners of the quarks and leptons is unrealistic because some of them are lighter than the usual quarks and leptons[27]. However, the potential is unstable at the tree level because some vacuum expectation values remain undertermined. It is not impossible that the effective potential resolves the mass splitting problem of the classical Lagrangian by the introduction of some higher scale[26] which would give radiatively a fermion-boson mass splitting at least of order M_W.

Finally, we mention that apart from conjectures of non-perturbative effects responsible for supersymmetry breaking[9] another possibility relies on explicit but soft supersymmetry breaking[27], which, although inelegant, would not spoil the ultra-violet improvement of supersymmetric theories[28].

5. Yang-Mills Interactions With Local Supersymmetry and Superhiggs Mechanism Without a Cosmological Constant

Nowadays supersymmetry seems a unique symmetry of quantum field theory which could give a consistent scheme for particle unification including gravity. In particular, supersymmetric Yang-Mills theories[18] could solve the hierarchy problem[10] of GUT's and explain the naturalness of the small scale of the standard model. When these theories are coupled to supergravity[29] it is tempting to believe that they could even solve the last hierarchy problem, namely the connection between the weak interaction and the Planck scale.

In the present lecture we report on a recent investigation[30] which led to the most general coupling of arbitrarily many chiral multiplets, transforming under some representation of a given gauge group G and interacting in a supersymmetric invariant way through Yang-Mills and gravitational forces. The mathematical problem to be solved therefore consists of the coupling of supersymmetric Yang-Mills theories to N = 1 supergravity.

After the discovery of supergravity[29] it was possible to couple pure supersymmetric Yang-Mills theories[31] to N = 1 supergravity throught the tedious Noether procedure. With the later discovery of a tensor calculus[32] for local supersymmetry it was also possible to obtain arbitrary supersymmetric Lagrangians and to study interesting physical problems such as the Higgs and superHiggs[33] effect in curved space. In particular, it was realized[34] that when a superchiral multiplet is "minimally" coupled to supergravity, a model-independent mass formula arises which connects the gravitino mass to the scalar partners of the would-be Goldstone fermion of global supersymmetry. In the presence of arbitrarily many scalar multiplets, this mass formula generalizes to a model-independent mass sum rule which raises the scalar masses with respect to the fermion masses. Among the consequences of this result, it has been argued[35] that a large supersymmetry breaking[36,30] should occur at a scale of $10^{10} - 10^{11}$ GeV.

I would now like to describe how the most general Lagrangian invariant under Yang-Mills transformations and local supersymmetry has been constructed[30]. Using superspace considerations, the action for coupling chiral multiplets to supergravity and to Yang-Mills fields is

$$\int d^4x\, d^4\theta\, E\,\left[\Phi(\bar{S}e^{2\tilde{g}V}, S) + \mathrm{Re}(1/R\; g(S))\right] \quad , \qquad (5.1)$$

where S_i are covariantly chiral scalar multiplets with components (z_i, χ_{Li}, h_i) and V is a G Lie algebra valued vector multiplet with components in the Wess-Zumino gauge given by $V^\alpha = (0,0,0,B^\alpha{}_m,\lambda^\alpha,D^\alpha)$. E is the superspace determinant. The gravitational multiplet has components[30]

$$e_{a\mu}\ ,\ \psi_{\mu\alpha}\ ,\ u\ ,\ A_m \quad .$$

The particle fields are z_i, χ_{Li} for the chiral multiplets, $B^\alpha{}_m$, λ^α for the vector multiplet and $e^a{}_\mu, \psi_\mu$ for the gravitational multiplet. The auxiliary components, which can eventually be eliminated through their algebraic field equations, are h_i, D^α, A_m and $u = S - iP$. The Yang-Mills part of the theory has the general form

$$\int d^4x\, d^4\theta\, E\, \mathrm{Re}\left[1/R\; f_{\alpha\beta}(S)\; W^\alpha{}_a\; \varepsilon^{ab}\; W^\beta{}_b\right] \quad , \qquad (5.2)$$

in terms of the superfield strength multiplet[18] $W^\alpha{}_a$ $(a = 1,2)$, and of an arbitrary function $f_{\alpha\beta}(S)$, transforming as the symmetric product of the adjoint representation of G. It has been assumed that the Lagrangian does not contain more than two derivatives in boson fields and one derivative in fermion fields. Only terms quadratic in the Yang-Mills fields have been retained. Also, a Fayet-Iliopoulos term for possible U(1) groups has not been considered. A Fayet-Iliopoulos term changes the present analysis because the superpotential g(z) must be R-symmetric[36] and because of additional compensating transformations[37]. Gauge invariance also implies the following conditions on the functions $\phi(z,z^*)$ and on the superpotential g(z)

$$z^{*j}\, T^\alpha{}_j{}^i\, \Phi'_i = z_j\, T^\alpha{}_i{}^j\, \Phi'^i \quad ; \quad z^{*j}\, T^\alpha{}_j{}^i\, g'_i = 0 \quad , \qquad (5.3)$$

where $\phi(z,z^*)$ is real, g(z) analytic and

$$\phi'_i = \frac{\partial\phi}{\partial z^{*i}} \quad , \quad \phi'^i = \frac{\partial\phi}{\partial z_i} \quad , \quad g'^i = \frac{\partial g}{\partial z_i} \quad , \quad g'_i = \frac{\partial g^*}{\partial z^{*i}} \quad .$$

By making a superWeyl transformation on the supervierbein, one can immediately realize that the final Lagrangian only depends on the real function

$$G(z,z^*) = 3 \log\left(-\frac{\phi(z,z^*)}{3}\right) - \log(|g(z)|^2/4) \tag{5.4}$$

and on the analytic function $f_{\alpha\beta}(z)$. In fact, $G''_i{}^j = \partial^2 G/\partial z^i \partial z^*{}_j$ and $\mathrm{Re}(f_{\alpha\beta}(z))$ act as metrics for the kinetic part of the scalar and vector fields, respectively. The action formula for a local chiral multiplet S of components z, χ_L, h and for a local vector multiplet of components C, s_L, H, B_a, χ_R, D are given by

$$e^{-1} L_S = h + uz + \overline{\psi}_R\cdot\gamma\, \chi_L + \overline{\psi}_R{}^\mu\, \sigma_{\mu\nu}\, \psi_R{}^\nu\, z$$

$$e^{-1} L_V = D - i/2\overline{\psi}\cdot\gamma\gamma_5\lambda - 1/3(Hu^* + H^*u) + 2/3\, B_m A^m - \tag{5.5}$$

$$-i/3\, \overline{S}\gamma_5\gamma\cdot R^P + i/4\, \varepsilon^{mnrs}\overline{\psi}_m\gamma_n\psi_r(B_s - 1/2\overline{\psi}_s S) - 2/3\, C\, L_{SG}\, e^{-1} \tag{5.6}$$

respectively. Here, L_{SG} is the supergravity action including the auxiliary fields and $R^P{}_\mu$ is the Poincaré covariant Rarita-Schwinger operator.

By means of the multiplication rules of chiral and vector multiplets, one uses Eq. (5) for the second term in Eq. (1) and for the term in Eq. (2), while one uses Eq. (6) for the first term in Eq. (1). The final result for the three terms in

$$e^{-1} L_\phi = -1/6\, \phi\, L_{SG}\, e^{-1} + \phi''^i{}_j(-1/2\, D_\mu z_i\, D^\mu z^{*j} - \overline{\chi}_{Li}\, \not{D}\chi_R{}^j$$
$$+ 1/2\, h_i\, h^{*j}) - \phi''{}'_k{}^{ij}\, \overline{\chi}_{Li}\, \chi_{Lj}\, h^{*k} + \phi''{}'_k{}^{ij}\, \overline{\chi}_{Li}\, \not{D}z_j\, \chi_R{}^k$$
$$+ 1/2\, \phi''''_{kl}{}^{ij}\, \overline{\chi}_{Li}\, \chi_{Lj}\, \overline{\chi}_R{}^k\, \chi_R{}^l + 1/3\, u^*(\phi'^i\, h_i$$
$$- \phi''^{ij}\, \overline{\chi}_{Li}\, \chi_{Lj}) + i/3\, A^\mu(1/2\, \phi''^i{}_j\, \overline{\chi}_R{}^j\, \gamma_\mu\, \chi_{Li} + \phi'^i(D_\mu z_i -$$
$$- \overline{\psi}_{\mu L}\, \chi_{Li})) + \phi''^i{}_j\, \overline{\psi}_{\mu L}\, \not{D}z^{*j}\, \gamma^\mu\, \chi_{Li} - 4/3\, \phi'^i\, \overline{\chi}_{Li}\, \sigma^{\mu\nu} D_\mu\, \psi_{\nu L}$$
$$- 1/8\, e^{-1}\, \varepsilon^{\mu\nu\rho\sigma}\, \overline{\psi}_\mu\, \gamma_\nu\, \psi_\rho(\phi'^i\, D_\sigma z_i + 1/2\, \phi''_i{}^j\, \overline{\chi}_R{}^i\, \gamma_\sigma\, \chi_{Lj}) -$$
$$- 1/2\, \phi''^i{}_j\, \overline{\psi}_{\mu L}\, \chi_{Li}\, \overline{\psi}_R{}^\mu\, \chi_R{}^j + 1/6\, \phi'^i\, \overline{\chi}_{Li}\, (\psi_{\mu L}\, \overline{\psi}\cdot\gamma\, \psi^\mu +$$
$$+ 1/2\, \sigma^{\mu\nu}\, \psi_L{}^\rho\, \overline{\psi}_\nu\gamma_\rho\, \psi_\mu + \sigma^{\mu\nu}\, \psi_{\nu L}\, \overline{\psi}_\mu\, \gamma\cdot\psi) +$$

$$+ 1/2\ \tilde{g}\ \phi'^{i}\ T^{\alpha}{}_{i}{}^{j}\ z_j(D^{\alpha} + i\overline{\psi}_L\cdot\gamma\ \lambda_R{}^{\alpha})$$

$$- 2i\tilde{g}\ \phi''{}_{i}{}^{j}\ T^{\alpha}{}_{j}{}^{k}\ z_k\ \overline{\lambda}_R{}^{\alpha}\ \chi_R{}^{i} + \text{h.c.} \tag{5.7}$$

$$e^{-1}\ L_g{}^{pot} = -\ 1/2\ g''^{ij}\ \overline{\chi}_{Li}\ \chi_{Lj} + 1/2\ g'^{i}\ h_i + 1/2\ g\ u$$

$$+ 1/2\ \overline{\psi}_R\cdot\gamma\ \chi_{Li}\ g'^{i} + 1/2\ g\ \overline{\psi}_{\mu R}\ \sigma^{\mu\nu}\ \psi_{\nu R} + \text{h.c.} \tag{5.8}$$

$$e^{-1}\ L_f{}^{YM} = 1/2\ f_{\alpha\beta}(-1/4\ F_{\mu\nu}{}^{\alpha}\ F^{\mu\nu\beta} - 1/2\ \overline{\lambda}^{\alpha}\ \tilde{\not{D}}\lambda^{\beta} + 1/2\ D^{\alpha}\ D^{\beta} +$$

$$+ 1/4\ F_{\mu\nu}{}^{\alpha}\ \overline{\psi}_{\rho}\ \sigma^{\mu\nu}\ \gamma^{\rho}\ \lambda^{\beta} + 1/4\ \tilde{F}_{\mu\nu}{}^{\alpha}\ F^{\mu\nu\beta} - 1/2\ D_{\mu}(\overline{\lambda}_L{}^{\alpha}\ \gamma^{\mu}\ \lambda_R{}^{\beta}))$$

$$+ 1/2\ f'^{i}{}_{\alpha\beta}(\overline{\chi}_{Li}\ (-\sigma\cdot\hat{F}^{\alpha} + iD^{\alpha})\lambda_L{}^{\beta} - 1/2\ h_i\ \overline{\lambda}_L{}^{\alpha}\ \lambda_L{}^{\beta} -$$

$$- 1/2\ \overline{\psi}_R\cdot\gamma\ \chi_{Li}\ \overline{\lambda}_L{}^{\alpha}\ \lambda_L{}^{\beta}) + 1/4\ f''^{ij}{}_{\alpha\beta}\ \overline{\chi}_{Li}\ \chi_{Lj}\ \overline{\lambda}_L{}^{\alpha}\ \lambda_L{}^{\beta} + \text{h.c.} \tag{5.9}$$

Moreover, a Weyl rescaling and local chiral rotations enable us to put the Einstein and Rarita-Schwinger terms in canonical form and, furthermore, to exhibit the dependence of the final action on $G(z,z^*)$ and $f_{\alpha\beta}(z)$

$$e \rightarrow \exp(4\sigma)\ e \quad ,$$

$$\lambda \rightarrow \exp(-3\sigma/2)\ \lambda \quad ,$$

(Weyl rescaling)

$$\chi_i \rightarrow \exp(-\sigma/2)\ \chi_i \tag{5.10}$$

$$\psi_{\mu} \rightarrow \exp(\sigma/2)\ (\psi_{\mu} + 2\gamma_{\mu}\ \sigma^{i}\ \chi_{Li} + 2\gamma_{\mu}\ \sigma_i\ \chi_R{}^{i})$$

$$\exp(2\sigma)\phi = -3 \quad ,$$

(Chiral transformations)

$$\psi_{\mu L} \rightarrow (g/g^*)^{1/4}\ \psi_{\mu L}$$

$$\chi_{Li} \rightarrow (g/g^*)^{-1/4}\ \chi_{Li} \tag{5.11}$$

$$\lambda_{Ri} \rightarrow (g/g^*)^{-1/4}\ \lambda_{Ri}$$

Inspection of Eqs. (7) and (8) shows that have the same structure as given by Cremmer et al[34] with trivial index contractions with the addition of the last three terms in Eq. (7) and with normal derivatives replaced by gauge covariant derivatives.

The final form of the potential[30] is

$$-1/4\ e^{-J}\left[\hat{g}^{i}\ \hat{g}^{*}_{j}\ (J''^{i}_{j})^{-1} + 3|g|^{2}\right] + 1/2\ \tilde{g}^{2}\ (J'^{i}\ T^{\alpha}{}_{i}{}^{j}\ z_{j})$$

$$(J'^{i}\ T^{\beta}{}_{i}{}^{j}\ z_{j})\mathrm{Re}(f_{\alpha\beta}{}^{-1}) = -e^{-G}\left[G'^{i}\ G'_{j}(G''^{i}_{j})^{-1} + 3\right]$$

$$+ 1/2\ \tilde{g}^{2}\ \mathrm{Re}(f_{\alpha\beta}{}^{-1})\ (G'^{i}\ T^{\alpha}{}_{i}{}^{j}\ z_{j})(G'^{i}\ T^{\beta}{}_{i}{}^{j}\ z_{j}) \qquad (5.12)$$

with $\hat{g}^{i} = g'^{i} - g\ J'^{i}$ and

$$G(z,z^{*}) = 3\ \log\ (-\ \phi/3) - \log(|g|^{2}/4) = J - \log(|g|^{2}/4) \quad .$$

This is due to the invariance of the system under the transformation

$$J(z,z^{*}) \rightarrow J(z,z^{*}) + f(z) + f^{*}(z^{*}) \quad , \qquad (5.13)$$

$$g(z) \rightarrow 2g(z)\ e^{f(z)} \quad .$$

The kinetic terms of the scalar fields and the vector fields are given by

$$e\ G''_{i}{}^{j}(z,z^{*})\ D_{\mu}z_{j}\ D_{\nu}z^{*i}\ g^{\mu\nu} \qquad (5.14)$$

and

$$-1/4\ e\ \mathrm{Re}(f_{\alpha\beta}(z))\ g^{\mu\rho}\ g^{\nu\sigma}\ F^{\alpha}{}_{\mu\nu}\ F^{\beta}{}_{\rho\sigma} \quad , \qquad (5.15)$$

respectively. We would now like to discuss the superHiggs effect and mass formulae for several chiral multiplets. Following Ref. [34], we consider the "minimal" coupling to supergravity. This we define by the two conditions

$$G''_{i}{}^{j}(z,z^{*}) = -1/2\ \delta_{i}{}^{j} \quad , \ f_{\alpha\beta}(z) = \delta_{\alpha\beta} \quad . \qquad (5.16)$$

The complete fermionic bilinear terms become

$$e^{-G/2}\left[\bar{\psi}_{R\mu}\ \sigma^{\mu\nu}\ \psi_{R\nu} - \bar{\psi}_{R}\cdot\gamma\ G'^{i}\ \chi_{Li} - \bar{\chi}_{Li}(G'^{i}\ G'^{j} - G''^{ij})\chi_{Lj}\right]$$

$$-i/4\ \tilde{g}\ z^{*j}\ T^{\alpha}{}_{j}{}^{i}\ z_{i}\ \bar{\psi}_{R}\cdot\gamma\ \lambda_{L}{}^{\alpha} + i\tilde{g}\ \bar{\lambda}_{L}{}^{\alpha}\ \chi_{Li}\ z^{*j}\ T^{\alpha}{}_{j}{}^{i} + \mathrm{h.c.}$$

(5.17)

The Goldstone fermion is defined by the coupling $-\bar{\psi}_{R}\cdot\gamma\ \eta_{L}$ and is

$$\eta_{L} = e^{-G/2}\ G'^{i}\ \chi_{Li} + i/4\ \tilde{g}\ z^{*j}\ T^{\alpha}{}_{j}{}^{i}\ z_{i}\ \lambda_{L}{}^{\alpha} \quad . \qquad (5.18)$$

The scalar potential for $\tilde{g} = 0$ is

$$V = e^{-G}\left[2G'_{i}\ G'^{i} - 3\right] \quad . \qquad (5.19)$$

If supersymmetry is broken, $e^{-G} \neq 0$ at the extremum. Moreover, the two conditions $V = 0$ (cancellation of the cosmological constant) and $V'_i = 0$ imply

$$G_i \, G'^i = 3/2 \quad , \quad G''_{ij} \, G'^j = 1/2 \, G'_i \quad . \tag{5.20}$$

If we now consider the mass matrix (17) and separate the Goldstone mode given by (18), we get, for the over-all spin 1/2 squared mass matrix

$$T_r \, (M_{1/2})^2 = 2 \, e^{-G}\left[4 \, G''_{ij} \, G''^{ij} - 1\right] \quad . \tag{5.21}$$

The contribution of the gravitino mass to the fermion mass matrix is

$$4 \, e^{-G} = 4 \, (m_{3/2})^2 \quad , \tag{5.22}$$

while the spin 0 squared mass matrix is

$$T_r \, (M_0)^2 = 4 \, e^{-G}\left[2 \, G''_{ij} \, G''^{ij} + N/2\right] \quad . \tag{5.23}$$

From (21), (22) and (23) we get

$$\begin{aligned} \text{Supertrace } M^2 &= T_r \, (M_0)^2 - T_r \, (M_{1/2})^2 - T_r \, (M_{3/2})^2 \\ &= 2 \, (N-1) \, (m_{3/2})^2 \, . \end{aligned} \tag{5.24}$$

Equation (24) shows that in supergravity the scalar masses are raised with respect to the fermion masses without the need of a Fayet-Iliopoulos mechanism. In terms of the supersymmetry breaking scale

$$m = V_o^{1/4} \quad , \tag{5.25}$$

the gravitino mass is

$$m_{3/2} = \frac{1}{\sqrt{3}} \frac{m^2}{\sqrt{\mu \, M_{Pl}}} \quad , \tag{5.26}$$

where μ is a mass of order 100 GeV. If $m \ll \sqrt{\mu \, M_{Planck}}$, the correction given by Eq. (24) is completely negligible. On the contrary, if $m \gtrsim \sqrt{\mu \, M_{Planck}}$ then $m_{3/2} \sim O(\mu)$ and the splitting due to supergravity couplings to the particle masses cannot be neglected.

This result shows that models based on large breaking of supersymmetry must be reconsidered when supergravity corrections are taken into account.

In order to show an explicit model in which the superHiggs

effect occurs, we consider an N + 1 component chiral multiplet z_i with superpotential

$$g(z)/2 = M^2(z_1 + \beta) \quad . \tag{5.27}$$

The analysis goes as in Ref. [34]. If we adjust the coefficient β to be $(2\sqrt{2}-\sqrt{6})M_{Planck}$, then an absolute minimum with zero cosmological constant occurs at $z_1 = (-\sqrt{2}+\sqrt{6})M_{Planck}$, $z_i = 0$ $i \neq 1$. The gravitino mass is

$$m_{3/2} = \frac{M^2}{M_{Pl}} \sqrt{2} \; e^{1/4(\sqrt{6}-\sqrt{2})^2} \quad .$$

The other particle masses are

$$m_{A_1}{}^2 = 2\sqrt{3}\, m_{3/2}{}^2 \quad , \qquad m_{B_1}{}^2 = 2(2-\sqrt{3})\, m_{3/2}{}^2 \quad ,$$

$$m_{A_i}{}^2 = m_{B_i}{}^2 = m_{3/2}{}^2 \quad , \qquad i = 2,\ldots,N+1 \quad .$$

$$m_{\chi i} = 0 \quad , \qquad i = 2,\ldots,N+1 \quad .$$

The would-be Goldstino m_{χ_1} has been eaten by the gravitino to provide the superHiggs effect. This example automatically extends to gauge interactions provided z_1 is a singlet and the other N-z_i's transform according to some non-trivial representation of G. This example shows that the mass term for the scalar fields $1/2\; z_i z^{*i}\, m^2{}_{3/2}$ acts on an explicit breaking of global supersymmetry. This term cannot be neglected if $m \sim \sqrt{\mu M_{Planck}}$, giving rise to a splitting which invalidates the mass relations of globally supersymmetric theories[20]. This is precisely what is needed to raise the masses of the scalar partners of quarks and leptons as shown in the example of Ref. [29]. A further remark[35] is that, as a consequence of Eq. (24), even the scalar fields which have a vacuum expectation value $v_i << M_{Planck}$ get masses of the same order as the gravitino mass. In view of the light Higgs sector of the theory this gives a gravitino mass $m_{3/2} \leqslant 100 - 1000$ GeV or equivalently $m \leqslant 10^{10} - 10^{11}$ GeV. Interestingly enough, cosmological considerations[38] exclude a supersymmetry breaking $m \geqslant 10^6$ GeV and smaller than $10^{10} - 10^{11}$ GeV. Therefore, the mass formula given by Eq. (24) seems to imply a unique scale of large supersymmetry breaking, namely $10^{10} - 10^{11}$ GeV [25]. In view of the fact that it is difficult to make a concrete model in which supersymmetry breaking occurs at $m \sim \mu$ it seems interesting to envisage the possibility that supergravity may predict a definite supersymmetry breaking scale and to solve the large hierarchy by the relation

$$m^2 \simeq \mu\, M_{Pl} \quad ,$$

where μ is the weak interaction scale[39].

One may wonder how general these conclusion are. Of course they depend on the fact that we demanded cancellation of the cosmological constant at the tree level - a requirement that could be relaxed. The conclusion could be invalidated by gauge interactions. In addition, as pointed out in Ref. [35], radiative corrections or dynamical effects may change the situation. Lastly, the result could perhaps be modified by taking the functions G and $f_{\alpha\beta}$ defined by Eqs. (1) and (2) to differ in their canonical values given by Eq. (16).

References

1. S. Glashow, Nucl. Phys. 22 (1961) 579;
   J.C. Ward and A. Salam, Phys. Lett. 13 (1964) 168;
   S. Weinberg. Phys. Rev. Lett. 19 (1967) 1264.
   A. Salam, "Elementary Particle Theory", ed. N. Svartholm, Alquimist and Wiksell, Stockholm, 1968, p. 367.
2. For reviews on supersymmetry see for instance:
   P. Fayet and S. Ferrara, Phys. Rep. 32C (1977) 251;
   A. Salam and J. Strathdee, Fortsch. Phys. 26 (1976) 57.
3. For a review see for instance:
   P. van Nieuwenhuizen, Phys. Rep. 68 (1981) 184.
4. R. Haag, J.T. Lopuszanski and M. Sohnius, Nucl. Phys. B88 (1975) 257.
5. S. Coleman and J. Mandula, Phys. Rev. 159 (1967) 1251;
   L.O'Raifeartaigh, Phys. Rev. Lett. 14 (1965) 575; Phys. Rev. B139 (1965) 1052.
6. P. Fayet, Nucl. Phys. B113 (1976) 135.
7. F. Gliozzi, J. Scherk and D. Olive, Nucl. Phys. B122 (1977) 253; L. Brink, J.H. Schwarz and J. Scherk, Nucl. Phys. B121 (1977) 11.
8. S. Ferrara and B. Zumino, unpublished;
   M. Sohnius and P. West, Phys. Lett. 100B (1981) 245;
   K.S. Stelle, LPTENS preprint 81/24 (1981).
9. L. Maiani in Proceedings of the Summer School of Gif-sur-Yvette (1979), p. 3;
   E. Witten, Nucl. Phys. B188 (1981) 313.
   S. Dimopoulos and S. Raby, Nucl. Phys. B192 (1981) 353.
10. E. Gildener and S. Weinberg, Phys. Rev. D13 (1976) 3333;
    S. Weinberg, Phys. Lett. 82B (1979) 387.
11. G. 't Hooft, Cargese Lectures 1979, to be published;
    M. Veltman, Acta Physica Polonica, to be published;
12. Y.A. Gol'fand and E.P. Likhtam, JETP Lett. 13 (1971) 323;
    D.V. Volkov and V.P. Akulov, Phys. Lett. 46B (1973) 109;
    J. Wess and B. Zumino, Nucl. Phys. B70 (1974) 39.
13. S. Ferrara, B. Zumino and J. Wess, Phys. Lett. 51B (1974) 239.
14. A. Salam and J. Strathdee, Nucl. Phys. B80 (1974) 499; Nucl. Phys. B84 (1975) 127; see also
    S. Ferrara, CERN preprint TH.2957 (1981), Plenary talk given at

the 9th International Conference on General Relativity and Gravitation, Jena (1980), to be published.
15. See the third Ref. in (12).
16. J. Wess and B. Zumino, Nucl. Phys. B78 (1974) 1.
17. A. Salam and J. Strathdee, Nucl. Phys. B76 (1974) 477
18) S. Ferrara and B. Zumino, Nucl. Phys. B79 (1974) 413;
A. Salam and J. Strathdee, Phys. Lett. 51B (1974) 353.
19. L. O'Raifeartaigh, Nucl. Phys. 56B (1975) 413; Nucl. Phys. B89 (1975) 41 - B96 (1975) 331.
20. S. Ferrara, L. Girardello and F. Palumbo, Phys. Rev. D20 (1979) 403.
21. J. Wess and B. Zumino, Phys. Lett 49B (1974) 52;
J. Iliopoulos and B. Zumino, Nucl. Phys. B76 (1974) 310;
S. Ferrara, J. Iliopoulos and B. Zumino, Nucl. Phys. B77 (1974) 413;
S. Ferrara and O. Piguet, Nucl. Phys. B93 (1975) 261.
22. W. Fishler, H. Nilles, J. Polchinski, S. Raby and L. Susskind, Phys. Rev. Lett. 47 (1981) 757.
23. P. Fayet in "Unification of the Fundamental Particle Interactions", ed. by S. Ferrara, J. Ellis and P. van Nieuwenhuizen (Plenum Press, N. Y., 1980), p. 587.
24. S. Weinberg, Harvard preprint HUTP 81/A047 (1981).
25. R. Barbieri, S. Ferrara and D.V. Nanopoulos, CERN preprint TH.3226 (1982).
26. E. Witten, Phys. Lett 105B (1981) 267
27. S. Dimopoulos and H. Georgi, Nucl. Phys. B193 (1981) 150.
28. L. Girardello and M.T. Grisaru, Brandeis University preprint (1981).
29. D.Z. Freedman, P. van Nieuwenhuizen and S. Ferrara, Phys. Rev. D13 (1976) 3214;
S. Deser and B. Zumino Phys. Lett. 62B (1976) 335.
For a recent review, see
P. van Nieuwenhuizen, Phys. Rep. 68 no. 4 (1981) 189.
30. E. Cremmer, S. Ferrara, L. Girardello and A. Van Proeyen, CERN preprint TH.3312 (1982) and CERN preprint TH.3348 (1982).
31. S. Ferrara, J. Scherck and P. van Nieuwenhuizen, Phys. Rev. Lett. 37 (1976) 1976;
S. Ferrara, L. Gliozzi, J. Scherk and P. van Nieuwenhuizen, Nucl. Phys. B117 (1976) 333;
32. S. Ferrara and P. van Nieuwenhuizen, Phys. Lett. 76B (1978) 404; K.S. Stelle and P.C. West, Phys. Lett. 77B (1978) 376.
33. D.V. Volkov and V.A. Soroka, JETP Lett. 18 (1973) 312;
S. Deser and B. Zumino, Phys. Rev. Lett. 38 (1977) 1433.
34. E. Cremmer, B. Julia, J. Scherk, S. Ferrara, L. Girardello and P. van Nieuwenhuizen, Phys. Lett. 79B (1978) 231; Nucl. Phys. B147 (1979) 105.
35. J. Ellis and D.V. Nanopoulos, CERN preprint TH.3319 (1982)
36. R. Barbieri, S. Ferrara, D.V. Nanopoulos and K.S. Stelle, CERN preprint TH. 3243 (1982), to appear in Phys. Lett. B.

37. K.S. Stelle and P.C. West, Nucl. Phys. B145 (1978) 175;
M. Sohnius and P.C. West, Phys. Lett. 105B (1981) 353.
38. H. Pagels and J. Primack, Phys. Rev. Lett. 48 (1982) 223;
S. Weinberg, Phys. Rev. Lett. 48 (1982) 1303.
39. For an attempt to use local supersymmetry in supercolour schemes in connection with the hierarchy problem, see:
H.P. Nilles, CERN preprint TH.3294 (1982).

SUPERGROUPS AND THEIR REPRESENTATIONS

I. Bars

J. W. Gibbs Laboratory
Department of Physics
Yale University
New Haven, Conn. 06520 U.S.A.

1. Applications of Supergroups and Superalgebras in Physics

About a decade ago a new kind of symmetry principle appeared in physics, namely supersymmetry. The novel feature of this symmetry is that it operates between bosons and fermions which have different space-time (or spin and statistics) properties. The generators of supersymmetry transformations form a Lie superalgebra whose even subalgebra is an ordinary Lie algebra and the odd generators, which mix bosons and fermions, close under anti-commutation to the even part.

The first applications of relativistic supersymmetry appeared in string theory[1] and field theory[2,3]. Recent applications of static supersymmetry occurred in nuclear physics[4]. Recently a general theory that uses induced representations of non-compact supergroups in supersymmetric field theory has also been developed[5]. A different kind of application of superalgebras with a reinterpretation of (odd-even) → (left-right of Lorentz group) has been used in composite models of quarks and leptons to obtain theories of dynamically unbroken $SU(N)_L \times SU(N)_R$ chiral symmetries[6]. Finally, supergroup invariant integration has been developed in the context of path integrals for the quantized super "chiral" SU(N/M) model analogous to the symmetric top problem[7]. There is one other early attempt which tries to use superalgebras for interval symmetries[8], but this approach is plagued by unphysical ghosts, for which so far only partial cures have been suggested. The cures are related to another application of superalgebras[9] in connection to the BRS symmetry and the Faddeev-Popov ghosts that appear in quantum gauge theories.

Most of these applications are to rather different problems and are unrelated to each other. We may, however, organize them according to the kinds of representations that they use: finite- and infinite-dimensional representations, as shown below.

I. <u>Compact Supergroups</u>
<u>Finite Dimensional Representations</u>

(1) Nuclear Physics (Static Supersymmetry)[4]
SU(6/M).

(2) Composite Quarks and Leptons[6]
SU(N + 4/N), SU(N/N) x SU(M/M).

(3) Super Chiral Model (≈ Symmetric top)[7].
SU(N/M).

(4) Internal Supergroups[8]
SU(N,1).

(5) BRS Symmetry in Quantum Gauge Theories[9].

II. <u>Non-Compact Supergroups</u>
<u>Infinite Dimensional Representations</u>

(1) a) Globally Supersymmetric Field Theories[2 3].

b) Supersymmetric Grand Unification[10].

c) Supersymmetric Preon Models[11]
(Super Poincaré group).

(2) Locally Supersymmetric Field Theory-Supergravity[12]
(Super Poincaré, OSp(N/4) , SU(N/4), etc.).

(3) Superstring Theory[1]
Super Kac-Moody.

(4) General Superfield Theory with Induced Representations[5]
Super Poincaré in d-dimensions,
OSp(N/2M), SU(N/M).

Among these applications only nuclear physics has provided experimental support for the usefulness of superalgebras in Nature. The theory is based on the superalgebra SU(6/M) in which the 3- dimensional rotation group which assigns spins to nuclear states is embedded. This provides a classification scheme for many low lying nuclear states of several even-even (bosonic) and even-odd (fermionic) nuclei in the platinum-gold region. The classification in a single irrep. of SU(6/M) predicts: (1) an energy formula for patterns of many nuclear levels, (2) relations among decay rates of excited states and (3) relations between nucleon-transfer reactions among such nuclei. The predictions work not only qualitatively but also quantitatively within 10-20%. The 10-20% discrepancies notwithstanding, the fact that a large amount of data on different kinds of experiments is successfully described by this simple scheme is in itself remarkable. It has been shown that the supersymmetry scheme can be extended to situations which are far more complex than the original application[4]. Superalgebras have thus provided a rather general approach for correlating and organizing nuclear data as well as making many testable predictions. The whole approach is still under experimental investigation and theoretical development.

Supersymmetric field theory is generally more convergent than non-supersymmetric field theory and there are hopes that some theories such as N=4 super Yang-Mills theory may even be finite to all orders of perturbation theory. The improvement of the convergence of field theory is a remarkable effect of supersymmetry, Unfortunately, there is no supersymmetric field theory in particle or gravitational physics which makes testable predictions of this kind of supersymmetry either directly or indirectly. There remains to be seen if more imaginative applications in field theory and particle physics will lead to measurable effects of supersymmetry. The formalism in ref. [5] is being developed to look for other possibilities.

Super string theory is one of the very early areas where superalgebras appear. However, strings are idealizations of what might happen in the theory of strong interactions, that is quantum chromodynamics (QCD). Strings have their own unphysical problems which are quite independent from the underlying field theory (QCD).

The use of superrepresentations in theories of composite quarks and leptons made from preons is very different[6]. The Lagrangians of these models are invariant under global internal symmetries $U(N)_L$ which transform left-handed fermions, and $U(M)_R$ which transform right-handed fermions. Furthermore, there is a pre-color (or hypercolor) local gauge symmetry which binds the preons. There is no supersymmetry of the usual kind (although the model could be enlarged to include them). A supergroup or superalgebra SU(N/M) is introduced which includes the symmetry of the Lagrangian $SU(N)_L$ x $SU(M)_R$ x U(1) as the even subgroup. However, the even-odd grading of the generators is not according to bose-fermi, but rather, according to left-right. So, the odd supergenerators (which are not symmetries) mix left with right. Even though SU(N/M) is not a symmetry, it plays the role of a classification group. This classification of preons predicts sets of composite quarks and leptons which solves[6] certain constraints[13] called "decoupling", which are necessary for the dynamical survival of the chiral symmetries $SU(N)_L \times SU(M)_R \times U(1)$. The survival of chiral symmetries is needed to explain the small masses of composite quarks and leptons.

In these supergroup models there is a larger supergroup (again not a symmetry) which has an even subgroup which includes precolor together with the chiral symmetries

$$\text{precolor} \times SU(N)_L \times SU(M)_R \times U(1)$$

The classification of anti-preons together with the composite quarks and leptons in the same supermultiplet of this supergroup (not a symmetry) solves another constraint, called "anomaly matching", also needed for the dynamical survival of the chiral symmetry. A few more constraints needed for the same purpose are uniquely and simultaneously solved provided we restrict ourselves to just the supergroups $SU(N + 4/N)$ in the two-box superrepresentation , and $SU(N/N) \times SU(M/M)$ in the two-box superrepresentation . This notation will be explained below.

The BRS symmetry of quantum gauge theories can be viewed in the context of superalgebras[9] . This symmetry is used to prove the unitarity of the theory by generating Ward identities that show the cancellation of all ghost effects.

The toy model of the $SU(N/M)$ chiral model in 1-time dimension (like the symmetric top) is developed[7] for the purpose of understanding the meaning of supergroup invariant integration. This model can easily be solved by ordinary canonical quantization of its bosons and fermions and any Green function can be obtained. On the other hand, by considering also the path integral quantization (bosonic and fermionic integration) one must consider supergroup invariant integration via a super Haar measure. Since we know all the answers already from the point of view of canonical quantization, the meaning and rules of supergroup integration can be developed, as in ref. [7].

In the rest of these lectures I will not return to the applications. The interested reader can read the literature. I will now concentrate on the mathematical aspects of the representations of supergroups and superalgebras as developed in refs. [14,15]. Some new results not found in refs. [14,15] will also appear in this paper. Further developments[16] on Kac-Dynkin diagrams and on noncompact supergroups and unitary representations[17] will appear elsewhere. Other approaches to representations of superalgebras can be found in the literature[18,19].

2. The Superalgebra SU(N/M)

Just as a complete classification of Lie algebras was obtained by Cartan, a complete classification of Lie superalgebras has been given by Kac[18]. It is useful to compare these two classifications and describe some analogies between the fundamental representations of Lie algebras and superalgebras.

Lie algebras fall into 4 infinite classes A_n, B_n, C_n, D_n with n = 1, 2, 3 ..., and five exceptional cases T_2, F_4, E_6, E_7, E_8. The 4 infinite classes can be put in one-to-one correspondence with the classical groups $A_{n-1} \longleftrightarrow SL(n)$ or $SU(n)$; $B_n \longleftrightarrow SO(2n+1)$; $C_n \longleftrightarrow Sp(2n)$; $D_n \longleftrightarrow SO(2n)$. Similarly, Lie superalgebras fall into 5 infinite classes $A(n,m) \longleftrightarrow SL(n+1/m+1)$; $B(n,m) \longleftrightarrow OSp(2n+1/2m)$; $D(n,m) \longleftrightarrow OSp(2n/2m)$; $P(n)$; $Q(n)$ and the exceptional cases $D_\alpha(2,1)$, $G(3)$, $F(4)$, plus the Cartan superalgebras. By comparing the structure of the matrices in the fundamental representation for Lie algebras and superalgebras, we can establish some useful analogies. Correspondences obtained in this way were used to develop the representation theory in terms of supertableaux[14,17].

In these lectures we concentrate only on SU(N/M), which is the compact form of SL(N/M). We will describe here the compact version and its unitary finite dimensional representations. A similar treatment was given for the finite dimensional representations of Osp(N/2M) and P(N)[14]. Unitary representations of noncompact SL(N/M) have also been constructed[17] but will not be discussed here.

In the N+M dimensional fundamental representation the algebra of SU(N/M) is given by hermitian (N+M) x (N+M) matrices of the form

$$H = \left[\begin{array}{c|c} H_N & \theta \\ \hline \theta^+ & H_M \end{array}\right] \quad , \quad \mathrm{Tr}\, H_N = \mathrm{Tr}\, H_M \quad . \tag{1}$$

Here H_N (H_M) is an NxN (MxM) hermitian matrix $H_N = H_N^+$ ($H_M = H_M^+$) constructed from bosonic complex parameters. That is, H_{11}, H_{12} etc. are complex bosonic numbers. On the other hand θ is an NxM matrix filled with complex fermions or complex anticommuting Grassmann numbers. That is, θ together with θ^+ describe 2NM real anticommuting Grassmann numbers. Even powers of Grassmann numbers will be included in the set of bosons, while odd powers will be included in the set of fermions. The traceless part of H_N (H_M) corresponds to an SU(N) (SU(M)) subgroup, while the trace part generates a U(1). Thus the bosonic part of H forms a Lie subalgebra corresponding to SU(N) x SU(M) x U(1).

As an example, consider SU(2/1) in the form

$$H = 1/2 \left[\begin{array}{cc|c} \multicolumn{2}{c|}{\multirow{2}{*}{$\vec{\omega}\cdot\vec{\tau} + \omega^4\,\tau_4$}} & \theta^1 - i\theta^2 \\ & & \theta^3 - i\theta^4 \\ \hline \theta^1 + i\theta^2 & \theta^3 + i\theta^4 & 2\omega^4 \end{array}\right] \tag{2}$$

$\vec{\tau}$ = Pauli matrices, τ_4 = unity .

This matrix contains 4 real bosonic parameters ω^1, ω^2, ω^3, ω^4 and 4 real fermionic parameters θ^1, θ^2, θ^3, θ^4. The generators in the fundamental representation are given as the matrix coefficients of these real variables:

$$\vec{T} = \left[\begin{array}{c|c} \vec{\tau}/2 & 0 \\ \hline 0 & 0 \end{array}\right] , \quad T_4 = \left[\begin{array}{cc|c} 1/2 & & \\ & 1/2 & \\ \hline & & 1 \end{array}\right]$$

$$S_1 = 1/2 \left[\begin{array}{cc|c} \multicolumn{2}{c|}{0} & 1 \\ & & 0 \\ \hline 1 & 0 & 0 \end{array}\right] , \quad S_2 = 1/2 \left[\begin{array}{cc|c} \multicolumn{2}{c|}{0} & -i \\ & & 0 \\ \hline i & 0 & 0 \end{array}\right] \tag{3}$$

$$S_3 = 1/2 \left[\begin{array}{cc|c} \multicolumn{2}{c|}{0} & 0 \\ & & 1 \\ \hline 0 & 1 & 0 \end{array}\right] , \quad S_4 = 1/2 \left[\begin{array}{cc|c} \multicolumn{2}{c|}{0} & 0 \\ & & -i \\ \hline 0 & i & 0 \end{array}\right]$$

where T_m, m = 1,2,3,4 are ordinary Lie generators forming an SU(2) x U(1) subgroup, while S_μ , μ = 1,2,3,4 are the supergenerators.

Note the similarities between the SU(2/1) and SU(3) generators. However, despite appearances they are totally different. First, the S_μ close into the $\vec{T}$, T_4 under anticommutation rather than commutation as can be verified by explicit matrix multiplication

$$\{S_\mu, S_\nu\} = C_{\mu\nu}{}^m\, T_m \quad . \tag{4}$$

Second, T_4 instead of being a traceless 3x3 matrix is a supertraceless 3x3 matrix. The supertrace of a matrix such as H is defined as

$$\mathrm{Str}\ H = \mathrm{Tr}\ H_N - \mathrm{Tr}\ H_M \tag{5}$$

and we have Str H = 0 ; Str T_4 = 0.

The analogy between SU(2/1) and SU(3), or more generally

SU(N/M) and SU(N+M) is improved by combining the generators together with the parameters into the form H. That is,

$$H = \sum_{n=1}^{4} \omega^m T_m + \sum_{\mu=1}^{4} \theta^\mu S_\mu = \left[\begin{array}{c|c} H_N & \theta \\ \hline \theta^+ & H_M \end{array}\right] . \qquad (6)$$

In this form we need to consider only commutators and never anticommutators. This is because the product $\theta^\mu S_\mu$ essentially acts like a bosonic generator. Namely, consider two sets of fermionic anticommuting parameters θ^μ and $\tilde{\theta}^\mu$ and multiply equation (4) by the bosonic product $\theta^\mu \tilde{\theta}^\nu$.

$$\theta^\mu \tilde{\theta}^\nu \{S_\mu, S_\nu\} = i\theta^\mu \tilde{\theta}^\nu \; C_{\mu\nu}{}^a \; T_a \quad . \qquad (7)$$

Noting the anticommutation property of the fermions we can write, by changing their orders,

$$\begin{aligned} \theta^\mu \tilde{\theta}^\nu \{S_\mu, S_\nu\} &= \theta^\mu \tilde{\theta}^\nu S_\mu S_\nu - \tilde{\theta}^\nu \theta^\mu S_\nu S_\mu \\ &= [\theta^\mu S_\mu, \tilde{\theta}^\nu S_\nu] \quad . \end{aligned} \qquad (8)$$

Therefore, we see that closure is obtained with commutators:

$$[\theta^\mu S_\mu, \tilde{\theta}^\nu S_\nu] = i\theta^\mu \tilde{\theta}^\nu f_{\mu\nu}{}^n T_n \sim i \left[\begin{array}{c|c} H_N & 0 \\ \hline 0 & H_M \end{array}\right] . \qquad (9)$$

From this one learns that, as long as the parameters are combined with the generators, we only need to consider commutators in order to close the algebra. However, one must not forget to respect the order in which Grassmann numbers appear in a product. In general, we find that for SU(N/M) we have the Lie commutation rule

$$[H, H'] = i H'' \qquad (10a)$$

or, more explicitly,

$$\left[\left[\begin{array}{c|c} H_N & \theta \\ \hline \theta^\dagger & H_M \end{array}\right] , \left[\begin{array}{c|c} H'_N & \theta' \\ \hline \theta^{\dagger\prime} & H'_M \end{array}\right] \right] = i \left[\begin{array}{c|c} H''_N & \theta'' \\ \hline \theta^{\dagger\prime\prime} & H''_M \end{array}\right] , \qquad (10b)$$

where H_N , H_M , θ'' , $\theta^{\dagger\prime\prime}$ can be calculated by explicit matrix multiplication.

$$\begin{aligned} i\theta'' &= H_N\theta' + \theta H'_M - H'_N\theta - \theta' H_M \quad , \\ i\theta '' &= -\theta^{\dagger\prime} H_N - H'_M\theta^\dagger + \theta^\dagger H'_N + H_M\theta^{\dagger\prime} \quad , \\ iH''_N &= [H_N , H'_N] + \theta\, \theta^{\dagger\prime} - \theta'\theta^\dagger \quad , \end{aligned} \qquad (11)$$

$$iH''_M = [H_M , H'_M] + \theta^+\theta' - \theta^{+\prime}\theta \quad .$$

It is easy to verify that tr H''_N = tr H''_M. In the limit $\theta = \theta' = 0$ we obtain the SU(N) x SU(M) x U(1) subgroup. In this form the analogy between SU(N/M) and SU(N+M) is complete. For SU(N+M) the U(1) generator in H is traceless rather than supertraceless and theta is bosonic rather than fermionic; otherwise the commutation rules look formally similar to SU(N/M). Note that for iH''_N to be an antihermitian NxN matrix we must define the operation of hermitian conjugation to change the orders of the Grassmann numbers in a product. For example

$$(\theta_1\theta_2^{+\prime})^+ = \theta_2'\theta_1^+ \quad . \tag{12}$$

This is consistent with a physicist's interpretation of a Grassmann number as being simply a fermion. This definition of hermitian conjugation is in fact necessary in order to apply supergroups in quantum theory.

3. Fermions and Grassmann numbers

At this point it may be useful to describe anticommuting Grassmann numbers in a physicist's language. It is nothing but an annihilation or creation operator of a fermion, but not both together. In this language it is also possible to obtain a matrix description of a Grassmann number. Consider the creation-annihilation operators of a single fermion, which satisfy

$$\{a, a\} = 0 \quad , \quad \{a^+, a^+\} = 0 \quad , \quad \{a, a^+\} = 1 \quad . \tag{13}$$

From these one can construct only two states in a Fock space which is a direct product with the space on which the generators of the superalgebra act.

$$|1\rangle = |0\rangle = \text{vacuum}$$

$$|2\rangle = a^+|0\rangle = \text{one-particle state} \quad .$$

In this 2-dimensional vector space, one can calculate the matrix elements of the operator a

$$\langle i|a|j\rangle \sim \begin{bmatrix} 0 & 1 \\ 0 & 0 \end{bmatrix}_{ij} \quad . \tag{14}$$

The operator a, or its matrix representation taken with matrix multiplication, has all of the properties of a Grassmann number. To construct 2 Grassmann numbers consider two sets of creation-annihilation operators a_1, a_1^+, a_2, a_2^+. One can now construct 4 states $|0\rangle$, $a_1^+|0\rangle$, $a_2^+|0\rangle$, $a_1^+a_2^+|0\rangle$ and obtain 4x4 matrix representations of the operators a_1, a_2. These two operators or their matrix representation have all the properties of 2 Grassmann numbers. Similarly with n sets of operators a_i, a_i^+, i=1,2,... one obtains $2^n \times 2^n$ representations of the n operators a_i which can be considered as matrix representations of Grassmann numbers. From this physicist's description you learn that you should manipulate Grassmann numbers just like fermions, mainly respecting their anticommutation properties. In general, as will become evident below, we will need and infinite "pool" of Grassmann numbers in order to describe all possible supergroup elements, just as we need an infinite "pool" of bosonic commuting numbers (or an infinite number of values for the bosonic parameters). You can do differentiation and integration with Grassmann numbers with the same rules as fermions. The differential operator is just the other half of the fermion, that is, the canonical conjugator to a_i, which are $a_i^+ \to \frac{\partial}{\partial a_i}$ and the integration rule is as in path integrals.

4. The Supergroup SU(N/M)

We have seen that the algebra of SU(N/M) closes with commutators as in Eq. (10) provided the generators are multiplied by the parameters. It is also true that in this form the ordinary Jacobi identities are satisfied trivially

$$[[H_1, H_2], H_3] + \text{cyclic } 1,2,3 = 0 \quad . \tag{15}$$

This means that we obtain a group element by exponentiating H or by considering an infinite product of infinitesimal transformations

$$U_A{}^B = (e^{iH})_A{}^B = \lim_{n\to\infty} [(1 + i \frac{H}{n})^n]_A{}^B \quad . \tag{16}$$

The set of group elements U satisfy the ordinary group properties under matrix multiplication

$$\begin{aligned} &U_1 U_2 = U_3 \quad , && \text{(closure)} \\ &(U_1U_2)U_3 = U_1(U_2U_3) \quad , && \text{(associativity)} \\ &U^+ = U^{-1} = e^{-iH} \quad . && \text{(inverse)} \end{aligned} \tag{17}$$

This is the supergroup SU(N/M). Because of the group property it is possible to consider it as an invariance group of transformations on the variables of a physical theory, just as ordinary Lie groups. In the SU(N/M) superalgebra we need 2NM real or NM complex fermionic parameters $\theta_a{}^\alpha, a = 1,2,\ldots,N;\ \alpha = 1,2,\ldots,M$ to describe a given general group element. In order to consider a different group element we need a different set of NM complex fermionic parameters θ', just as we would need a different set of bosonic parameters. Therefore, we should not limit the pool of available anticommuting parameters to just 2NM real fermions. In general, just as we need ordinary bosonic variables which can take an infinite number of values we also need an infinite pool of anticommuting variables or fermions (which could be described by infinite dimensional matrices). The 2NM fermionic variables in any given SU(N/M) supergroup element can be picked from this infinite set. We can then use differentiation and integration rules for fermions, whenever needed, according to the same rules as in quantum theory and path integrals.

5. The Fundamental Basis or Module

Taking a set of states, "wavefunctions", or operators that transform like the basis states in the fundamental representation (i.e., a module) we can write the transformation rule

$$\Phi_A \rightarrow \Phi'_A = U_A{}^B\, \Phi_B \quad . \tag{18}$$

Such a basis will be denoted by the supertableau $\boxslash$ in analogy to a Young tableau. These states can be arranged in a column matrix in the form

$$\Phi_A = \begin{bmatrix} \phi_a \\ \\ \psi_\alpha \end{bmatrix} \quad \begin{matrix} a = 1,2,\ldots,\ N \quad , \\ \\ \alpha = 1,2,\ldots,\ M \quad , \end{matrix} \tag{19}$$

such that Φ_a are bosons and ψ_α are fermions. Such a wavefunction will be said to belong to class I fundamental representation. A second wavefunction $\tilde{\Phi}_A$ can be introduced, such that

$$\tilde{\Phi}_A = \begin{bmatrix} \psi_a \\ \\ \phi_\alpha \end{bmatrix} \quad \begin{matrix} a = 1,2,\ldots,\ N \quad , \\ \\ \alpha = 1,2,\ldots,\ M \quad , \end{matrix} \tag{20}$$

where ψ_a = fermion, and ϕ_α = boson. $\tilde{\Phi}$ will be said to belong to class II. Mostly, we will deal with class I because class II is obtained from class I by a "tilde" operation which commutes with the group element, as we will see.

Taking into account the fermionic property of θ, the transformed variables $\Phi'_A = (\phi'_a, \psi'_\alpha)$ preserve their bosonic or fermionic properties. With an infinitesimal transformation

$$U_A{}^B \approx \delta_A{}^B + i\, H_A{}^B \approx \left(e^{iH}\right)_A{}^B \tag{21}$$

we can write

$$\delta\phi_a = (H_N)_a{}^b\ \phi_b + \theta_a{}^\beta\ \psi_\beta = \text{boson} \quad ,$$

$$\delta\psi_\alpha = (\theta^+)_\alpha{}^b\ \phi_b + (H_M)_\alpha{}^\beta\ \psi_\beta = \text{fermion} \quad , \tag{22}$$

where we note that the product of two fermions $\theta\psi$ is considered a boson, while the product of a boson and a fermion $\theta^+\phi$ or $H_M\psi$ is considered a fermion. Hence the concept of boson includes even powers of fermions while the concept of fermion includes odd powers of fermions.

Under the SU(N) x SU(M) x U(1) subgroup (i.e., $\theta = \theta^+ = 0$) it is clear that

$$\phi_a \sim (\square\ , 1)_{1/N} \rightarrow \text{fundamental representation of SU(N)} \quad ,$$

$$\psi_\alpha \sim (1\ , \square)_{1/M} \rightarrow \text{fundamental representation of SU(M)} \quad . \tag{23}$$

The U(1) charges are associated with the supertraceless U(1) generator

$$T_o \sim \left[\begin{array}{c|c} 1/N & 0 \\ \hline 0 & 1/M \end{array}\right] \quad . \tag{24}$$

It is useful to define the grade of the index $A = (a, \alpha)$. This is given by

$$g(A) \begin{cases} \nearrow g(a) = 0 & \text{if} \quad \phi_a = \text{boson} \quad , \\ \searrow g(\alpha) = 1 & \text{if} \quad \psi_\alpha = \text{fermion} \quad . \end{cases} \tag{25}$$

(The value of the grade is opposite for class II when ψ_a = fermion, ϕ_α = boson). We can use the grade to interchange the orders of indices or wavefunctions taking into account the bosonic and fermionic property of various terms

$$\Phi_A\ \Phi_B = (-1)^{g(A)g(B)}\ \Phi_B\ \Phi_A \quad ,$$

$$\Phi_A\ U_C{}^D = (-1)^{g(A)\ [g(C)-g(D)]}\ U_C{}^D\ \Phi_A \quad , \tag{26}$$

$$U_A{}^B\ U_C{}^D = (-1)^{[g(A)-g(B)][g(C)-g(D)]}\ U_C{}^D\ U_A{}^B \quad ,$$

$$H_{1A}{}^B\ H_{2C}{}^D = (-1)^{[g(A)-g(B)][g(C)-g(D)]}\ H_{2C}{}^D\ H_{1A}{}^B \quad , \qquad \text{etc.}$$

6. Supertrace and Superdeterminant

A useful concept is the supertrace, as defined above, which can be written as

$$\text{Str } U = \sum_A (-1)^{g(A)} U_A{}^A \quad ,$$
$$\text{Str } H = \sum_A (-1)^{g(A)} H_A{}^A = \text{Tr } H_N - \text{Tr } H_M \quad . \tag{27}$$

The supertrace has the cyclic property, as can be verified by using Eqs. (26)

$$\text{Str } H_1 H_2 = \text{Str } H_2 H_1$$
$$\text{Str } H_1 H_2 H_3 = \text{Str } H_3 H_1 H_2 = \text{Str } H_2 H_3 H_1 \text{ , etc.} \tag{28}$$

This is not true for the trace. Furthermore, the supertrace is invariant under similarity transformations

$$H_A{}^B \to U_A{}^{A'} H_{A'}{}^{B'} (U^{-1})_{B'}{}^B \tag{29}$$

Using the cyclic property we can show easily

$$\text{Str } U_1 H U_1^{-1} = \text{Str } U_1^{-1} U_1 H = \text{Str } H \quad ,$$
$$\text{Str } U_1 U U_1^{-1} = \text{Str } U_1^{-1} U_1 \; U = \text{Str } U \quad . \tag{30}$$

The superdeterminant can be defined via the supertrace in analogy to the definition of determinant

$$\text{Sdet } U = \exp(\text{Str } \ln U) \quad . \tag{31}$$

If $U = e^{iH}$, then Sdet U = 1 implies

$$\text{Str } H = 0 \quad . \tag{32}$$

The superdeterminant satisfies the factorization property

$$\text{Sdet } U_1 U_2 = \text{Sdet } U_1 \text{ Sdet } U_2 \tag{33}$$

and is invariant under similarity transformations $U \to U_1 U U_1^{-1}$. For graded matrices of the form $\left[\begin{array}{c|c} A & 0 \\ \hline 0 & B \end{array}\right]$ the superdeterminant reduces to a ratio of two ordinary determinants

$$\mathrm{Sdet}\left[\begin{array}{c|c} A & 0 \\ \hline 0 & B \end{array}\right] = \det A/\det B \quad . \tag{34}$$

Therefore Sdet U is not a polynomial in the matrix elements of U, and could not be written by using an analog of the completely antisymmetric Levi-Civita symbol.

7. The λ Matrices and Killing Metric

Just as we identified the matrix representation of the generators for SU(2/1), as in Eq. (3), it is useful to identify a set of hermitian λ-matrices corresponding to the generators of SU(N/M) as the coefficients of bosonic and fermionic parameters

$$(H)_A{}^B = (\lambda_I/2)_A{}^B\, w^I \quad , \tag{35}$$

where ω^I describes bosons as well as fermions, in the adjoint representation, depending on the index I. It is clear that the bosonic generators are block diagonal while the fermionic ones are block off-diagonal. They can chosen as in Eq. (3)

$$\mathrm{SU(N)}:\ 1/2\left[\begin{array}{c|c} \lambda_n & 0 \\ \hline 0 & 0 \end{array}\right], \quad \mathrm{SU(M)}:\ 1/2\left[\begin{array}{c|c} 0 & 0 \\ \hline 0 & \lambda_m \end{array}\right],$$

$$\mathrm{U(1)}:\ 1/2\sqrt{\frac{2NM}{N-M}}\left[\begin{array}{c|c} 1/N & 0 \\ \hline 0 & 1/M \end{array}\right] \tag{36}$$

Supergenerators:

$$S_1 = 1/2\left[\begin{array}{c|c} 0 & \begin{matrix}10\ldots \\ 00\ldots \\ \ldots\ldots\end{matrix} \\ \hline \begin{matrix}100\ldots \\ 000\ldots \\ \ldots\ldots \\ \ldots\ldots\end{matrix} & 0 \end{array}\right], \quad S_2 = 1/2\left[\begin{array}{c|c} 0 & \begin{matrix}-i0\ldots \\ 00\ldots \\ \ldots\ldots\end{matrix} \\ \hline \begin{matrix}-i0\ldots \\ 00\ldots \\ \ldots\ldots \\ \ldots\ldots\end{matrix} & 0 \end{array}\right]$$

etc.

where λ_n (λ_m) are the SU(N) (SU(M)) traceless λ-matrices of Gell-Mann chosen to be orthonormal

$$\mathrm{Tr}\ \lambda_m\lambda_{m'} = 2\ \delta_{mm'} \ \text{etc.} \tag{37}$$

Note that for SU(N/M) we exclude the U(1) generator since it becomes proportional to unity and commutes with all other matrices. Hence we do not worry about the normalization involving N-M in the denominator which is used only for N ≠ M.

Using the fact that $\mathrm{Str}(H_1 H_2)$ is invariant under similarity transformations (which induce the supergroup transformations on the adjoint representation) we can define the super Killing metric of the group as

$$2\, g_{IJ} = \mathrm{Str}\ \lambda_I\, \lambda_J \ , \tag{38}$$

which gives the invariant in the adjoint representation

$$\mathrm{Str}\ H_1\ H_2 = 1/2\ \omega_1^I\ \omega_2^J\ g_{IJ} \ , \tag{39}$$

where the sum over I,J runs over bosonic and fermionic indices. We note that when both I and J are bosonic indices g_{IJ} is symmetric but when both I and J are fermionic indices g_{IJ} is antisymmetric. That is,

$$g_{IJ} = g_{JI}\ (-1)^{g(I)g(J)} \ . \tag{40}$$

Explicitly, we find from Eqs. (36-38)

$$g_{IJ} = \begin{array}{c} \mathrm{SU(N)} \\[1.5em] \mathrm{U(1)} \\[1em] \mathrm{SU(M)} \\[2em] \mathrm{Super} \end{array} \left[\begin{array}{ccc|c|ccc|cccc} 1 & & & & & & & & & & \\ & 1 & & & & & & & & & \\ & & 1 & & & & & & & & \\ \hline & & & \pm 1 & & & & & \multicolumn{3}{c}{0} \\ \hline & & & & -1 & & & & & & \\ & & & & & -1 & & & & & \\ \hline & & & & & & & 0\ i & & & \\ & & & & & & & -i\ 0 & & & \\ & & & & & & & & 0\ i & & \\ & \multicolumn{3}{c}{0} & & & & & -i\ 0 & & \\ & & & & & & & & & \ddots & \\ & & & & & & & & & & \ddots \\ \end{array} \right] \ . \tag{41}$$

The U(1) entry is (+1) for N-M>0, (-1) for N-M < 0 and omitted for N-M = 0. In a similar way we can define covariant tensors with the indices of the adjoint representation by using the invariance of $\mathrm{Str}\ (H_1 H_2 \ldots H_n)$. In particular, considering $\mathrm{Str}\ (H^n)$ and factoring out $\omega^{I_1}\ \omega^{I_2}\ldots\omega^{I_n}$ we define the covariant tensor

$$d^{(n)}{}_{I_1 I_2 \cdots I_n} = \frac{1}{n!} \, \mathrm{Str}\left[\lambda_{I_1} \lambda_{I_2} \cdots \lambda_{I_n} \pm \text{permutations}\right] \quad . \tag{42}$$

The $\pm$ signs are given as $(-1)^{g(I_1)g(I_2)}$ etc., depending on the permutation. Then, under the interchange of any set of indices $d^{(n)}{}_{I_1 I_2 \cdots I_n}$ is either symmetric or antisymmetric depending on the bosonic-fermionic property of the indices. For n = 3 this is the generalization of the d_{abc} coefficients of SU(N).

The metric g_{IJ} and its inverse g^{JK} with property

$$g_{IJ}\, g^{JK} = \delta_I{}^K = g^{KJ}\, g_{JI} \tag{43}$$

can be used in order to write invariants of the group. For example, if the generators of the group are denoted by X_I (e.g., represented by $\lambda_{I/2}$) then the quadratic Casimir invariant is obviously

$$\begin{aligned} C_2 &= X_I\, g^{IJ}\, X_J \\ &\equiv X_I\, X^I = X^I\, X_I\, (-1)^{g(I)} \end{aligned} \tag{44}$$

where we have defined

$$X^I \equiv g^{IJ}\, X_J = X_J\, g^{IJ} = X_J\, g^{JI}\, (-1)^{g(I)g(J)} \tag{45}$$

which follows from Eq. (40) and the form of g_{IJ} in Eq. (41).

The λ_I matrices listed in Eq. (36), excluding the identity matrix, form a complete set in terms of which any hermitian super traceless matrix H can be expanded. The completeness relation (or super "Fierz" transformation) is written in terms of $\lambda_I\, g^{IJ}\, \lambda_J \equiv \lambda_I \lambda^I$ and can be verified to be

$$N \neq M:\ (\lambda_I)_A{}^B\, (\lambda^I)_C{}^D = 2\delta_A{}^D\, \delta_C{}^B\, (-1)^{g(C)g(B)} - \frac{2}{N-M}\, \delta_A{}^B\, \delta_C{}^D \quad . \tag{46}$$

In the notation of Eq. (36), the sum in the left-hand side of Eq. (46) has the form

$$\left[\begin{array}{c|c} \lambda_n & 0 \\ \hline 0 & 0 \end{array}\right]_A^{\ B} \left[\begin{array}{c|c} \lambda_n & 0 \\ \hline 0 & 0 \end{array}\right]_C^{\ D} - \left[\begin{array}{c|c} 0 & 0 \\ \hline 0 & \lambda_m \end{array}\right]_A^{\ B} \left[\begin{array}{c|c} 0 & 0 \\ \hline 0 & \lambda_m \end{array}\right]_C^{\ D}$$

$$+\frac{2NM}{N-M}\left[\begin{array}{c|c}1/N & \\ \hline & 1/M\end{array}\right]_A^{\ B}\left[\begin{array}{c|c}1/M & \\ \hline & 1/N\end{array}\right]_C^{\ D} + i\,(S_1)_A{}^B(S_2)_C{}^D$$

$$- i\,(S_2)_A{}^B\,(S_1)_C{}^D + \text{ etc.} \quad , \tag{47}$$

showing the origin of the N-M in the denominator in Eq. (46). When N = M the U(1) generator is omitted from the sum and then the completeness relation becomes

$$N = M:\ (\lambda_I)_A{}^B(\lambda^I)_C{}^D = 2\ \delta_A{}^D\delta_C{}^B(-1)^{g(C)g(B)}$$
$$-\ 1/N\ \delta_A{}^B\delta_C{}^D\left[(-1)^{g(A)g(B)} + (-1)^{g(C)g(D)}\right] \quad . \tag{48}$$

The second term in this expression can be arrived at by direct computation or by moving the U(1) generator in Eq. (47) to the right-hand side of Eq. (46) and then taking the limit as $M \to N$.

From Eqs. (44) and (46) we compute the matrix elements of the quadratic Casimir operator when the generators X_I are in the fundamental representation $X_I = \lambda_I/2$ $(N \neq M)$

$$(C_2)_A{}^D = (\lambda_I/2)_A{}^B\ (\lambda^I/2)_B{}^D \tag{49}$$
$$= 1/2\ \left[(N - M)^2 - 1\right]/(N-M) \quad \delta_A{}^D$$

and is found proportional to the identity, as expected.

When N = M, SU(N/M) does not exist in the fundamental representation, and its smallest representation is the adjoint[19]. This is because the matrix $\mathbb{I}$, which is not one of the generators, is reproduced in the (anti-) commutation relations of the SU(N/N) λ-matrices. That is, $[H, H']$ contains the matrix $\mathbb{I}$, and therefore the correct commutation rules for SU(N/N) are not represented by the fundamental λ-matrices. In fact, if we attempt to calculate $(\lambda_I\lambda^I)_A{}^D$ from Eq. (48), we find that it is not proportional to the identity, indicating that this is not a casimir invariant. We may however define an algebra for the SU(N/N) λ-matrices, by subtracting the part proportional to the identity. We introduce the starbracket

$$[H, H']^* \equiv [H, H'] - 1/2N\ \mathrm{tr}[H, H']\ \times \mathbb{I} \equiv i\ H'' \quad , \tag{50}$$

where we have used the ordinary trace, not the supertrace, since $\mathbb{I}$ is supertraceless for SU(N/N). The star-bracket $[H, H']^* = i\ H''$ gives the correct commutation rules for the generators, and produces the correct structure constants, but it is no longer just a Lie commutator of the fundamental matrices. In the adjoint and higher dimensional irreps, it will, of course, reduce to ordinary matrix commutation. Because of these complications we will ignore SU(N/N) in these lectures.

8. Complex Conjugate Representation

We may define the complex or hermitian conjugate of the wave-functions ϕ_A with an upper (contravariant) index as

$$\phi^{+}{}_{A} \equiv \phi^{A} \quad .$$

The transformation property in this basis is obtained by taking the hermitian conjugate of Eq. (18) and remembering to change the orders of the fermions in the product, as part of the definition of hermitian conjugation (see argument before Eq. (12)). Then we obtain

$$\phi^{A} \rightarrow \phi'^{A} = \phi^{B}\,(U^{+})_{B}{}^{A} \quad , \tag{51}$$

with U appearing on the right. This contravariant basis will be denoted by a supertableau with a dot

$$\phi^{A} \sim \dot{\boxslash} \quad . \tag{52}$$

Unlike the group SU(N), in which the contravariant basis is equivalent to an (N-1) dimensional antisymmetric tensor, in SU(N/M) the contravariant basis cannot be written in terms of the covariant one. This is because there is no invariant tensor analogous to the Levi-Civita completely antisymmetric tensor. This difference between SU(N) and SU(N/M) is related to the fact that the superdeterminant is not a polynomial (see Eq. (34)).

9. Harmonic Oscillator Representation

We introduce a set of bosonic and fermionic harmonic oscillators with the indices of the fundamental representation

$$\boxslash \sim \xi_A = \begin{bmatrix} b_a \\ f_\alpha \end{bmatrix} , \quad \dot{\boxslash} \sim \xi^{+A} = (b^{+a} \;\; f^{+\alpha}) \quad , \tag{53}$$

with commutation-anticommutation rules

$$[b_a,\, b^{+c}] = \delta_a{}^{c}, \quad \{f_\alpha,\, f^{+\beta}\} = \delta_\alpha{}^{\beta} \quad , \tag{54}$$

all other obvious (anti)-commutators being zero. Equivalently, we may write

$$[\xi_A,\, \xi^{+B}\} = \delta_A{}^{B} \quad , \tag{55}$$

where the graded commutator $[\cdot,\cdot\}$ is defined as

$$\xi_A\, \xi^{+B} - (-1)^{g(A)g(B)}\, \xi^{+B}\xi_A = [\xi_A, \xi^{+B}\} = \delta_A{}^B \quad . \tag{56}$$

We may combine these oscillators with $H_A{}^B$ in order to form quantum generators X_I as

$$\xi^{+A}\, H_A{}^B\, \xi_B \equiv X_I\, \omega^I \quad , \tag{57}$$

where (up to some minus signs coming from anti-commuting fermions f_α with the fermions ω^I) we obtain

$$X_I = (\lambda_I/2)_A{}^B\, \xi_B\, \xi^{+A}\, (-1)^{g(A)} = \mathrm{Str}\,(\xi\, \xi^+\, \lambda_I/2) \quad . \tag{58}$$

However, since we always obtain simplifications by combining parameters with generators we will always use the forms

$$\xi^+ H\, \xi \equiv X \,, \quad \xi^+ H'\xi = X' \quad , \quad \text{etc.} \tag{59}$$

Note that the anticommuting fermions appearing in $H_A{}^B$ must be taken to anticommute with the fermionic oscillators as well, because an infinitesimal transformation of the oscillators under the group, as in Eq. (22), must maintain the commutation rules of the oscillators. In fact, a supergroup transformation with $U_A{}^B$ on the oscillators, as in Eq. (18), is a canonical transformation because it leaves the commutation rules of Eq. (55) invariant. Now, consider commuting two generators by using the quantum commutators of Eqs. (54-56). We find a cancellation of minus signs leading to the simple result

$$\begin{aligned} [X, X'] &= [\xi^+ H\, \xi \;,\; \xi^+ H'\xi] \\ &= \xi^+[H, H']\, \xi \\ &= i\, \xi^+H''\xi = i\, X'' \quad , \end{aligned} \tag{60}$$

where H" has the same form as H or H' as demonstrated in Eq. (11) by using only commutators. We see that the quantum generators X (including parameters) close again under <u>commutation and not anticommutation</u>.

We may introduce many sets of creation-annihilation operators by adjoining an extra index on ξ_{Ai}, $i = 1,2,\ldots,k$. To save indices we will use a vector notation $\vec{\xi}_A$ instead of the index i. The generators formed from each ξ_{Ai} close among themselves for each i. Therefore, their sums defined as

$$X = \vec{\xi}^+ H\vec{\xi} \,, \; X' = \vec{\xi}^+ H'\vec{\xi} \quad , \tag{61}$$

(where the vector index i is summed over) will also close just as above

$$[X, X'] = i X'' \quad . \tag{62}$$

It is useful to consider the quantum operator group element $\hat{U}$ obtained by exponentiating X:

$$\hat{U} = \exp(iX) = \exp(i \vec{\xi}^{+} H \vec{\xi}) \quad . \tag{63}$$

From Eq. (60) we learn that these operators form a group

$$\hat{U}\,\hat{U}' = \hat{U}'' \quad .$$

Using the quantum commutators of Eq. (55, 56) we derive by explicit quantum commutation

$$[X, \vec{\xi}_A] = -H_A{}^B \vec{\xi}_B \ , \ [X, \vec{\xi}^{+A}] = \vec{\xi}^{+B} H_B{}^A \quad , \tag{64}$$

where the order of H and ξ are important to avoid extra (-1) signs. Furthermore, it follows inmediately

$$\begin{aligned} \hat{U}^{+} \vec{\xi}_A \hat{U} &= (e^{iH})_A{}^B \vec{\xi}_B \sim \boxslash \quad , \\ \hat{U}^{+} \vec{\xi}^{+A} \hat{U} &= \vec{\xi}^{+B} (e^{-iH})_B{}^A \sim \boxslash \quad , \\ \hat{U}^{+} (\vec{\xi}^{+} H' \vec{\xi}) \hat{U} &= \vec{\xi}^{+} [e^{-iH} H' e^{iH}] \vec{\xi} \quad . \end{aligned} \tag{65}$$

We define the covariant generators

$$\begin{aligned} \underline{X}_A{}^B &\equiv X_I (\lambda^I/2)_A{}^B \quad , \\ \mathrm{Str}\,(\underline{X}\, H) &= X \quad . \end{aligned} \tag{66}$$

By using the completeness of the λ's we can write

$$\underline{X}_A{}^B = \vec{\xi}_A \cdot \vec{\xi}^{+B} - \delta_A{}^B (\vec{\xi}^{+} \cdot \vec{\xi} + 1) \quad . \tag{67}$$

Then we note that

$$\hat{U}^{+} \underline{X}_A{}^B \hat{U} = (e^{iH})_A{}^{A'} \underline{X}_{A'}{}^{B'} (e^{-iH})_{B'}{}^B \quad , \tag{68}$$

and therefore

$$\sum_A \underline{X}_A{}^B \underline{X}_B{}^A (-1)^{g(A)} = \mathrm{Str}(\underline{X}^2) \tag{69}$$

is an invariant. The quantum quadratic Casimir operator is now given by

$$\hat{C}_2 = X_I\, X^I = 1/2\ \mathrm{Str}(\underline{\overline{X}}^2) \quad . \tag{70}$$

This can be rewritten in the form

$$\hat{C}_2 = 1/2\ \left[(\xi^+{}_i\cdot\xi_j)(\xi^+{}_j\cdot\xi_i) + (N-M-K)(\xi^+{}_i\cdot\xi_i) - 1/N-M\ (\xi^+{}_i\cdot\xi_i)^2\right] \tag{71}$$

where i, j = 1,2,...K run through the copies of oscillators. The dot products

$$\xi^+{}_i\cdot\xi_j \equiv \xi^+{}_i{}^A\ \xi_{jA} \tag{72}$$

are invariant under a group transformation according to

$$\hat{U}\ (\xi\ {}_i\cdot\xi_j)\hat{U} = \xi^+{}_i\ e^{-iH}\ e^{iH}\ \xi_j = \xi^+{}_i\cdot\xi_j \quad . \tag{73}$$

This is another way of seeing that the quadratic Casimir operator indeed commutes with $\hat{U}$ or with all the generators X, for any set of parmeters H.

If we specialize the only one set of oscillators i = j = k = 1 then $\hat{C}_2$ takes a simple form

$$\hat{C}_2 = \frac{N-M-1}{2(N-M)}\ \hat{n}\ (N-M+\hat{n}) \tag{74}$$

where $\hat{n}$ is the number operator which counts bosons plus fermions

$$\begin{aligned} \hat{n} &= \xi^+\cdot\xi = \hat{n}_b + \hat{n}_f \quad , \\ \hat{n}_b &= b^+\cdot b \quad , \\ \hat{n}_f &= f^+\cdot f \quad . \end{aligned} \tag{75}$$

Thus, $\hat{n}$ rather than $\hat{C}_2$ can be taken as the Casimir invariant. Note that even though $\hat{C}_2$ could not be defined for N = M, the invariant operator $\hat{n}$ certainly exists and is meaningful for SU(N/N).

The n'th higher order quantum Casimir operators can be constructed by using the covariant tensor $\underline{\overline{X}}_A{}^B$ of Eq. (66). Thus $\mathrm{Str}(\underline{\overline{X}}^p)$ is an invariant, and we can write the p'th order Casimir invariant as

$$\hat{C}_p = \mathrm{Str}(\underline{\overline{X}}^p) = X_{I_1}X_{I_2}\cdots X_{I_p}\ \mathrm{Str}(\lambda^{I_1}\lambda^{I_2}\cdots\lambda^{I_p}) \quad , \tag{76}$$

where the index in λ^I is raised by using the killing metric as in Eq. (45). In certain applications it is more convenient to define another p'th order Casimir operator by using the covariant tensors of Eq. (42)

$$\hat{C}'_p = X_{I_1} X_{I_2} \cdots X_{I_p} \, d^{(p)I_1 I_2 \cdots I_p} \quad , \tag{77}$$

where the indices are raised by using g^{IJ}. The $\hat{C}'_p$ are linear combinations of $\hat{C}_p$ and lower order casimirs of type $\hat{C}_{p-1}, \hat{C}_{p-2}, \cdots$, as can be seen by using $[X_{I_1}, X_{I_2}\} \sim X_{I_3}$ to change the orders in the various terms of $d^{(p)I_1 I_2 \cdots I_p}$. In the special case of a single set of creation-annihilation operators ξ_i, ξ^+_j $i = j = k = 1$ all of these casimir operators become a function of only the number operator,

$$\hat{n} = \xi^{+A} \xi_A = \hat{n}_b + \hat{n}_f$$

as can easily be seen from the form $Str(\underline{X}^p)$. Thus in the case of a single set of harmonic oscillators there is only one independent Casimir invariant, namely the number operator $\hat{n}$. It is clear that in order to construct the most general representation one must take many copies of oscillators.

The harmonic oscillators could also be assigned to the complex conjugate representation

$$\boxtimes \sim \eta^A = (c^a \ g^\alpha) \quad ; \qquad \boxtimes \sim \eta^+_A = \begin{bmatrix} c^+_a \\ g^+_\alpha \end{bmatrix} \tag{78}$$

where c^a is a boson and g^α is a fermion. These differ from Eq. (53) in the assignment of lower and upper indices. They satisfy the quantum rules

$$[\eta^+_A, \eta^B\} = -\delta_A{}^B (-1)^{g(A)} \tag{79}$$

as opposed to Eq. (55), in which the order is important. We can now define generators as in Eq. (57)

$$\eta^A (-H)_A{}^B \eta^+_B \equiv Y_I \omega^I \equiv Y \tag{80}$$

and find the Y_I satisfy the same commutation rules as the X_I, that is

$$\begin{aligned} [\eta(-H)\eta^+ , \quad \eta(-H')\eta^+] &= \eta(-[H, H'])\eta^+ \\ &= i \, \eta(-H'')\eta^+ \, . \end{aligned} \tag{81}$$

Therefore, by taking many copies of ξ's and η's we may introduce a total generator G_I constructed as

$$G_I = X_I + Y_I = Str(\vec{\xi} \cdot \vec{\xi}^+ \lambda_I/2) - Str(\vec{\eta}^+ \cdot \vec{\eta} \, \lambda_I/2) \quad ,$$

$$G = \vec{\xi}^{+}H\vec{\xi} + \vec{\eta}(-H)\vec{\eta}^{+} \quad , \tag{82}$$

$$G_A{}^B = G_I(\lambda^I)_A{}^B = \vec{\xi}_A \cdot \vec{\xi}^{+B} - \eta^{+}{}_A \cdot \eta^B - \frac{\delta_A{}^B}{N-M}\left(\vec{\xi}^{+} \cdot \vec{\xi} - \vec{\eta} \cdot \vec{\eta}^{+}\right) \quad .$$

Similarly, defining a total group operator that induces group transformations on all quantum oscillators

$$\hat{U} = e^{iG} = \exp i\left\{\vec{\xi}^{+}H\vec{\xi} + \vec{\eta}(-H)\vec{\eta}^{+}\right\} \quad , \tag{83}$$

we compute, by using the quantum commutators, as in Eq. (65), for any oscillator $\vec{\xi}_A$:

$$\hat{U}^{+}\vec{\xi}_A\hat{U} = (e^{iH})_A{}^B\vec{\xi}_B \sim \boxslash \quad ,$$

$$\hat{U}^{+}\eta^{+}{}_A\hat{U} = (e^{iH})_A{}^B\eta^{+}{}_B \sim \boxslash \quad ,$$

$$\hat{U}^{+}\xi^{+A}\hat{U} = \xi^{+B}(e^{-iH})_B{}^A \sim \bar{\boxslash} \quad ,$$

$$\hat{U}^{+}\eta^A\hat{U} = \eta^B(e^{-iH})_B{}^A \sim \bar{\boxslash} \quad , \tag{84}$$

$$\hat{U}^{+}G_A{}^B\hat{U} = (e^{iH})_A{}^{A'}G_{A'}{}^{B'}(e^{-iH})_{B'}{}^B \sim \bar{\boxslash}\boxslash \quad \text{(adjoint)} \quad ,$$

as expected. The orders of the factors are important, otherwise a multitude of minus signs will appear.

In the presence of both $\vec{\xi}$ and $\vec{\eta}$ the Casimir operators are constructed as

$$\hat{C}_p = \mathrm{Str}(G^p) \quad . \tag{85}$$

They clearly commute with all G_I, since $\hat{U}^{+}\hat{C}_p\hat{U} = \hat{C}_p$ as a result of Eq. (84).

10. The Tilde Operator

There is a further generalization concerning the introduction of oscillators in a class II basis. We will distinguish them from class I with a tilde sign

$$\boxslash \sim \tilde{\rho}_A = \begin{bmatrix} v_a \\ r_\alpha \end{bmatrix} ; \qquad \bar{\boxslash} \sim \tilde{\rho}^{+A} = (v^{+a} \ \ r^{+\alpha}) \quad , \tag{86}$$

with v = fermion, r = boson, as distinguished from Eq. (58).

$$\{v_a, v^{+b}\} = \delta_a{}^b \qquad [r_\alpha, r^{+\beta}] = \delta_\alpha{}^\beta \quad . \tag{87}$$

These can be combined to the same from as Eq. (55)

$$[\tilde{\rho}_A, \tilde{\rho}^{+B}\} = \delta_A{}^B \qquad (88)$$

provided the grades of the indices are interpreted properly. The class II oscillators transform under the group just in the same way as class I oscillators.

Out of this set we may again construct generators by combining with $H_A{}^B$

$$\tilde{X} = \tilde{\rho}^+ H \tilde{\rho} \quad , \quad \tilde{X}' = \tilde{\rho}^+ H'\rho, \text{ etc.} \quad ; \quad [\tilde{X}, \tilde{X}'] = i\,\tilde{X}'' \qquad (89)$$

in analogy to Eq. (60). Then, taking $U = \exp iX$ we have

$$\hat{U}^+ \tilde{\rho}_A \hat{U} = (e^{iH})_A{}^B \tilde{\rho}_B \quad \sim \boxslash \qquad (90)$$

etc., as in Eqs. (65,84), as expected.

Similarly, we also have class II oscillators in a contravariant basis

$$\boxslash \sim \tilde{\sigma}^A = (w^a \; s^\alpha) \quad ; \quad \boxslash \sim \tilde{\sigma}^+{}_A = \begin{bmatrix} w^+{}_a \\ s^+{}_\alpha \end{bmatrix} \quad , \qquad (91)$$

where w is a fermion and s is a boson. The generators

$$\tilde{Y} = \tilde{\sigma}(-H)\tilde{\sigma}^+ \qquad (92)$$

will have the usual properties.

The total quantum generator constructed from all oscillators in Eqs. (82), (89) and (92) is

$$Q = \vec{\xi}^+ H \vec{\xi} - \vec{\eta} H \vec{\eta}^+ + \vec{\tilde{\rho}}^+ H \vec{\tilde{\rho}} - \vec{\tilde{\sigma}} H \vec{\tilde{\sigma}}^+ , \qquad (93)$$

where we have included multiple copies of the oscillators through the vector notation $\vec{\xi}$ etc. Then the full quantum operator group element which acts on all oscillators is

$$U = e^{iQ} \quad . \qquad (94)$$

It satisfies equations similar to (84), including the class II oscillators. Casimir invariants in the full space are constructed from Q.

Now we are ready for the <u>tilde operator</u> t. Consider the N+M dimensional matrix

$$t = \left[\begin{array}{cccc|cccc} \theta_0 & & & & & & & \\ & \theta_0 & & & & & & \\ & & \ddots & & & & & \\ & & & \theta_0 & & & & \\ \hline & & & & -\theta_0 & & & \\ & & & & & -\theta_0 & & \\ & & & & & & \ddots & \\ & & & & & & & -\theta_0 \end{array}\right] , \tag{95}$$

$$(\theta_0)^2 = 1 \qquad \longleftrightarrow \qquad t^2 = 1 \quad ,$$

where θ_0, considered as a fermionic parameter, anticommutes with all the Grassmann numbers $\theta_a{}^\alpha$ in $H_A{}^B$ as well as with all the quantum fermionic oscillators. We will however take $(\theta_0)^2 = 1$ implying that θ_0 is the sum of a fermionic creation and annihilation operator in the language of Grassmann numbers described before. This makes θ_0 a Clifford number, constructed from a Grassmann number plus its derivative $\theta_0 = \theta + \partial/\partial\theta$; $(\theta_0)^2 = 1$. We emphasize that θ_0 with these properties is just a convenience but not necessary for the final point to be made below. If θ_0 is omitted from the argument a combination of commutators and anticommutators will have to be used as usual. The matrix t is fermionic; its supertrace is not zero, and therefore is not included in H.

By direct matrix multiplication, and the use of the anticommutation property of Grassmann numbers, it can be seen that t and H commute!

$$[t, H] = 0 \quad . \tag{96}$$

We may now construct the quantum tilde operator t as

$$\hat{t} = \vec{\xi}^{+} t \,\vec{\tilde{\rho}} + \vec{\tilde{\rho}}^{+} t \,\vec{\xi} - \vec{\eta}\, t \,\vec{\tilde{\sigma}}^{+} - \vec{\tilde{\sigma}}\, t \,\vec{\eta}^{+} . \tag{97}$$

Using Eq. (96) and the quantum oscillator commutation rules introduced before we find

$$\begin{aligned} \left[\hat{t}, \vec{\xi}_A\right] &= -\left(t\vec{\tilde{\rho}}\right)_A = (-1)^{g(A)} \vec{\tilde{\rho}}_A \,\theta_0 \quad , \\ \left[\hat{t}, \vec{\tilde{\rho}}_A\right] &= -\left(t\vec{\xi}\right)_A = -(-1)^{g(A)} \xi_A \theta_0 \quad , \text{ etc.} \quad , \end{aligned} \tag{98}$$

such that

$$[\hat{t}, Q] = 0 \quad , \tag{99}$$

implying that the operator $\hat{t}$ is a group invariant. We now construct the unitary transformation

$$\hat{T} = e^{i\hat{t}\,\pi/2} \quad ,$$
$$\hat{T} \begin{bmatrix} \xi \\ \rho \end{bmatrix} T = \begin{bmatrix} 0 & it \\ it & 0 \end{bmatrix} \begin{bmatrix} \xi \\ \rho \end{bmatrix} , \tag{100}$$

where the indices are suppressed. A similar equation holds for η and $\tilde{\sigma}$. This transformation interchanges class I with class II oscillators wherever they appear and as a result of Eq. (99) commutes with the group transformation $\hat{U} = \exp iQ$ which acts on the indices of the oscillators. Therefore, class I and class II representations of any dimension will transform in identical ways under the group transformation. When the tilda operation $\hat{t}$ is applied on all the states in a representation it will simply switch all bosons into fermions and viceversa.

$$\text{boson} \overset{\hat{t}}{\longleftrightarrow} \text{fermion}$$

and will insert appropriate minus signs as required by a representation of the matrix of Eq. (95). The new set of states thus obtained (fermions + bosons) have identical transformation properties as the old set (bosons + fermions) under the action of the group, because of Eq. (99).

Because of the existence of the operator $\hat{t}$ it will be sufficient to construct group elements on pure class I bases, since the group element on class II or mixed[14] class I + class II bases will always be the same as pure class I. (The mixed tensor products of class I + class II bases often yield states whose bose-fermi content is switched relative to pure class I or pure class II. For example, you can obtain a basis for an adjoint representation whose bose-fermi content is opposite to the generators. Some examples are found in Ref. [14]).

11. Higher Representations from Tensor Products and Supertableaux

Let us consider the transformation properties of the direct product of two wavefunctions following the rule of Eq. (18)

$$\Phi^{(1)}{}_A\, \Phi^{(2)}{}_B \to \Phi'^{(1)}{}_A\, \Phi'^{(2)}{}_B = U_A{}^{A'}\, \Phi^{(1)}{}_{A'}\, U_B{}^{B'}\, \Phi^{(2)}{}_{B'} \quad . \tag{101a}$$

Note that the order of factors could not be changed without introducing some minus signs, as in Eq. (26).

$$\Phi'^{(1)}{}_A\ \Phi'^{(2)}{}_B = (-1)^{g(A')[g(B)-g(B')]} U_A{}^{A'} U_B{}^{B'}\ \Phi^{(1)}{}_{A'}\ \Phi^{(2)}{}_{B'} \tag{101b}$$

In terms of supertableaux this direct product can be written as

The direct product is reducible. To obtain irreducible tensors we supersymmetrize, and introduce the supertableau notation

 $\sim\ \Phi^{(+)}{}_{AB} = \Phi^{(1)}{}_A\ \Phi^{(2)}{}_B + \Phi^{(2)}{}_A\ \Phi^{(1)}{}_B\ ,$

(102)

 $\sim\ \Phi^{(-)}{}_{AB} = \Phi^{(1)}{}_A\ \Phi^{(2)}{}_B - \Phi^{(2)}{}_A\ \Phi^{(1)}{}_B\ .$

Note the supertableau rule that the indices A, B are kept in the same order, but the wavefunctions $\Phi^{(1)}{}_A$, $\Phi^{(2)}{}_B$ are interchanged. This is necessary so that both terms in the sum transform in the same way with the overall factors

$$(-1)^{g(A')[g(B)-g(B')]}\ U_A{}^{A'}\ U_B{}^{B'}\ , \tag{103}$$

as in Eq. (101). Then the symmetric (antisymmetric) sum stays symmetric (antisymmetric) after the supergroup transformation. If we wish to keep the order of wavefunctions the same, we may use Eq. (26) to write also

 $\sim\ \Phi^{(+)}{}_{AB} = \Phi^{(1)}{}_A\ \Phi^{(2)}{}_B + (-1)^{g(A)g(B)}\ \Phi^{(1)}{}_B\ \Phi^{(2)}{}_A\ ,$

(104)

 $\sim\ \Phi^{(-)}{}_{AB} = \Phi^{(1)}{}_A\ \Phi^{(2)}{}_B - (-1)^{g(A)g(B)}\ \Phi^{(1)}{}_B\ \Phi^{(2)}{}_A\ .$

These arrangements of indices transform irreducibly by the above argument. Thus, under the interchange of indices we have the supersymmetric or superantisymmetric tensors

 $\sim\ \Phi^{(+)}{}_{AB} = (-1)^{g(A)g(B)}\ \Phi^{(-)}{}_{BA}\ ,$

(105)

 $\sim\ \Phi^{(-)}{}_{AB} = -\ (-1)^{g(A)g(B)}\ \Phi^{(-)}{}_{BA}\ .$

By specializing the indices $A = (a,\ \alpha)$, i.e. $\Phi_A = \begin{bmatrix}\phi_a \\ \psi_\alpha\end{bmatrix} \equiv (\phi_a\ ,\ \psi_\alpha)^T$ to bosons and fermions we identify the various components

$$\boxed{\diagup}\boxed{\diagup} \sim \Phi^{(+)}{}_{AB} = \begin{cases} \phi_{ab} = \phi_{ba} \sim (\square\square,\ 1)_{2/N} \ ; \ 1/2\ N(N+1) \text{ bosons}\ , \\ \phi_{a\beta} = \phi_{\beta a} \sim (\ \square , \square\ )_{1/N\ +\ 1/M} \ ; \ NM \text{ fermions}\ , \\ \phi_{\alpha\beta} = -\phi_{\beta\alpha} \sim (1, \begin{smallmatrix}\square\\\square\end{smallmatrix}\ )_{2/M} \ ; \ 1/2\ M(M-1) \text{ bosons}\ , \end{cases} \tag{106}$$

$$\begin{smallmatrix}\boxed{\diagup}\\\boxed{\diagup}\end{smallmatrix} \sim \Phi^{(-)}{}_{AB} = \begin{cases} \phi_{ab} = -\ \phi_{ba} \sim (\begin{smallmatrix}\square\\\square\end{smallmatrix},\ 1)_{2/N} \sim 1/2\ N(N-1) \text{ bosons}\ , \\ \phi_{a\beta} = -\phi_{\beta a} \sim (\square , \square\ )_{1/N\ +\ 1/M} \sim NM \text{ fermions}\ , \\ \phi_{\alpha\beta} = +\phi_{\beta\alpha} \sim (1, \square\square\ )_{2/M} \sim 1/2\ M(M+1) \text{ bosons}\ . \end{cases} \tag{107}$$

The tensors with an odd number of α indices are fermions by construction. However, we may construct class II tensors $\tilde{\Phi}^{(+)}{}_{AB}$, $\tilde{\Phi}^{(-)}{}_{AB}$ by substituting in Eq. (104) the class II first-rank tensor $\tilde{\Phi}^{(2)}{}_B = (\psi_b\ ,\ \phi_\beta)^T$ instead of the class I $\Phi^{(2)}{}_B = (\phi_b\ ,\ \psi_\beta)^T$ (keeping $\Phi^{(1)}{}_A$ the same) where these two differ in the nature of anticommuting variables (all ϕ's bosons and all ψ's fermions). Then the tilda transformed tensors $\tilde{\Phi}_{AB}$ will have boson or fermion components switched relative to the old ones

$$\tilde{\Phi}^{(+)}{}_{AB} = \begin{matrix} (\square\square\ ,\ 1) & = \text{fermions}\ , \\ (\square , \square\ ) & = \text{bosons}\ , \\ (1,\ \begin{smallmatrix}\square\\\square\end{smallmatrix}\ ) & = \text{fermions}\ , \end{matrix} \tag{108}$$

and similarly for $\tilde{\Phi}^{(-)}{}_{AB}$. Even though the Bose $\leftarrow\rightarrow$ Fermi properties are interchanged, Φ_{AB} and $\tilde{\Phi}_{AB}$ transform identically under the group. This is a manifestation analogous to

$$[\hat{t},\ Q] = 0 \ , \tag{109}$$

that was described in the last section in terms of the quantum oscillators. Because we can always construct class II bases from those of class I via the tilde operation, it will be sufficient to concentrate only on class I.

To obtain the matrix representation of the group element we write the transformation of the new basis as

$$\Phi'^{(+)}{}_{AB} = U_{AB}{}^{A'B'} \quad \Phi^{(+)}{}_{A'B'} \tag{110}$$

and extract a properly symmetrized group element in this higher dimensional representation (⧄⧄)

$$U_{AB}{}^{A'B'} = 1/2\ \Big\{(-1)^{g(A')[g(B)-g(B')]}\ U_A{}^{A'} U_B{}^{B'} + (-1)^{g(A)g(B)}\ (-1)^{g(A')[g(A)-g(B')]}\ U_B{}^{A'}\ U_A{}^{B'}\Big\} \quad , \tag{111}$$

which satisfies the superpermutation properties indicated by the supertableau

$$⧄⧄ : \quad U_{AB}{}^{A'B'} = (-1)^{g(A)g(B)}\ U_{BA}{}^{A'B'} = (-1)^{g(A')g(B')}\ U_{AB}{}^{B'A'} \quad . \tag{112}$$

Before we go on, let us also construct states in a Fock superspace which have the same tranformation properties as the tensors above. Let us start with the class I oscillators of Eq. (53). Define the vacuum state

$$\xi_A |0\rangle = 0 \quad , \quad \langle 0| \xi^{+A} = 0 \quad . \tag{113}$$

Then the one particle bra state

$$\langle 0| \xi_A \equiv \langle A| \tag{114}$$

transforms as the fundamental basis ⧄

$$\langle A| \rightarrow \langle A|\hat{U} = \langle 0|\hat{U}\ \hat{U}^{+}\xi_A\ \hat{U} \quad , \\ = \langle 0|(e^{iH})_A{}^B\ \xi_B \quad , \\ = (e^{iH})_A{}^B\ \langle B| \quad . \tag{115}$$

Similarly the two particle state

$$\langle A\ B| \equiv \langle 0| \xi_A \xi_B \tag{116}$$

is automatically supersymmetric, since the oscillators satisfy

$$\xi_A \xi_B = (-1)^{g(A)g(B)}\ \xi_B \xi_A \tag{117}$$

and therefore it transforms as ⧄⧄ , with the matrix of Eq. (111):

$$\langle AB|\ \hat{U} = U_{AB}{}^{A'B'}\ \langle A'B'| \quad . \tag{118}$$

We may in fact relate the two approaches of wavefunctions and Fock superspace states, by defining

$$\Phi^{(1)} = \Phi^{(1)}{}_A\ \xi^{+A} \quad , \quad \Phi^{(2)} = \Phi^{(2)}{}_A\ \xi^{+A} \quad , \quad \text{etc.} \tag{119}$$

and applying them on the vacuum -

$$\begin{aligned} &|\Phi^{(1)}\rangle \equiv \Phi^{(1)}\ |0\rangle \quad , \quad \text{etc.} \\ &|\Phi^{(1)}\ \Phi^{(2)}\rangle \equiv \Phi^{(1)}\ \Phi^{(2)}\ |0\rangle \quad , \quad \text{etc.} \end{aligned} \tag{120}$$

Then, it is clear that the wavefunctions can be defined as

$$\Phi^{(1)}{}_A = \langle A|\Phi^{(1)}\rangle \sim$$ 

(121)

$$\Phi^{(+)}{}_{AB} = \langle A\ B\ |\ \Phi^{(1)}\cdot\Phi^{(2)}\rangle \sim$$ 

and they will have the same transformation properties as the states in the Fock superspace.

This method of obtaining higher dimensional super tensors is obviously generalized

n

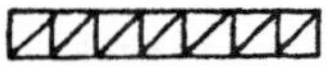

$$\sim \langle 0\ |\ \xi_{A_1}\ \xi_{A_2}\ \cdots\ \xi_{A_n} = \langle A_1 A_2 \cdots\ A_n\ | \ ,$$

(122)

$$\Phi^{(+)}{}_{A_1 A_2\ \ldots\ A_n} = \langle A_1 A_2 \cdots\ A_n\ |\ \Phi^{(1)}\ \Phi^{(2)}\ \ldots\ \Phi^{(n)}\rangle$$

To make supertableaux with more than one row we must use more than one set of creation-annihilation operators. In fact with K sets

$$\vec{\xi}_A = (\xi^{(1)}{}_A\ ,\ \xi^{(2)}{}_A\ ,\ \ldots\ ,\ \xi^{(K)}{}_A) \quad , \text{etc.}$$

we can construct up to K rows. Actually, with more than one set the method becomes cumbersome and we may instead simply use products of wavefunctions as in Eq. (102). For example, the tensor corresponding to the supertableau  is

1 2

3 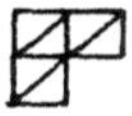

$$\begin{aligned} \Phi_{AB,C} = {} & \Phi^{(1)}{}_A\ \Phi^{(2)}{}_B\ \Phi^{(3)}{}_C + \Phi^{(2)}{}_A\ \Phi^{(1)}{}_B\ \Phi^{(3)}{}_C \\ & - \Phi^{(3)}{}_A\ \Phi^{(2)}{}_B\ \Phi^{(1)}{}_C - \Phi^{(2)}{}_A\ \Phi^{(3)}{}_B\ \Phi^{(1)}{}_C \quad . \end{aligned} \tag{123}$$

Note again the supertableau rule that the indices keep the same order and the wavefunctions are symmetrized-antisymmetrized according to the tableau. Note further that, as in usual Young tableaux, first the symmetrization is done for every row and next the antisymmetrization is done for every column. In the above example the net tensor has simple supersymmetry properties in the interchange of indices A ←--→ B but not in A ←--→ C or B ←--→ C, as in usual Young tableaux.

Matrix elements of the supergroup in this higher representation are now obtained by applying a transformation on every $\Phi^{(1)}{}_A$, $\Phi^{(2)}{}_B$, $\Phi^{(3)}{}_C$, and writing the result as

$$\Phi'_{AB,C} = U_{AB,C}{}^{A'B',C'}\ \Phi_{A'B',C'} \quad , \tag{124}$$

analogously to Eq. (110). It is left as an exercise to obtain these matrix elements in terms of the fundamental $U_A{}^B$ (see Ref. [14]).

There are more representations that involve the direct products of covariant ◩ with contravariant ◪ bases. For example, the adjoint representation ◪◩ which we have already encountered in Eq. (84) is such an example. As already emphasized (see Section 8), unlike ordinary (SU(N)) the contravariant indices ◪ could not be reproduced by taking direct products of covariant ones. The only thing to notice is that for a tensor

$$\Phi_{A_1A_2\ \dots\ A_n}{}^{B_1B_2\dots B_m} \tag{125}$$

to be irreducible it must satisfy the following conditions:

(1) The lower indices must be supersymmetrized according to a supertableau.

(2) The upper indices must independently be supersymmetrized according to some other supertableau.

(3) The tensor must be supertraceless in all pairs of upper-lower indices

$$\sum_{A_1} (-1)^{g(A_1)}\ \Phi_{A_1A_2\ \dots\ A_n}{}^{A_1B_2\dots B_m} = 0 \quad , \quad \text{etc.} \tag{126}$$

These rules are summarized by the following general supertableau:

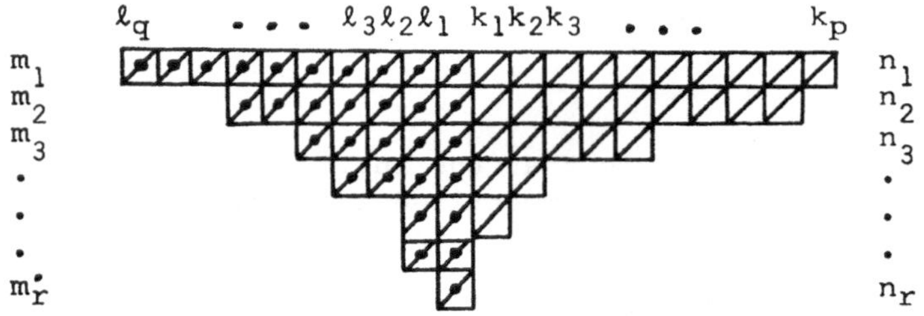

Fig. 1 Supertableau for SU(N/M) irreducible tensors, eq. (125)

$$n_1 \geq n_2 \geq \ldots n_r \geq 0 \quad ,$$

$$m_1 \geq m_2 \geq \ldots m_{\dot{r}} \geq 0 \quad .$$

The dotted boxes correspond to the upper indices and are drawn as mirror images of the usual Young tableaux. The same rules could be applied to SU(N) and it will give an irreducible tensor. If one wishes, one can transform the SU(N) dotted tableau to an undotted one by the simple replacement of every column with k-dotted boxes, by a column with N-k undotted boxes:

k  N-k

true for SU(N) ,

(false for SU(N/M)) .

This property, which is based on the Levi-Civita tensor, does not apply for supertableaux, since no such invariant tensor exists.

After obtaining an irreducible tensor, we arrive at the matrix representation of the group elements

$$U^{(A_1'A_2' \ldots A_n')(B_1B_2\ldots B_m)}_{(A_1A_2\ldots A_n)(B_1'B_2' \ldots B_m')} \quad , \tag{127}$$

by making a transformation on every ϕ_A and rearranging the order of the factors as in Eqs. (101b, 111). Further examples of these can be found in Ref. [14].

12. The Invariant Characters

We have already seen in Eq. (30) that the supertrace of the fundamental group element $U_A{}^B$ is invariant under similarity transformations with other group elements

$$\mathrm{Str}\, U_1 \, U \, U_1^+ = \mathrm{Str}\, U \quad . \tag{128}$$

This is the invariant character in the fundamental representation

$$\chi_{\square}(U) = \mathrm{Str}\, U \quad . \tag{129}$$

To construct the invariant characters in the higher dimensional representations we simply contract lower and upper indices of the matrix representations of the previous section according to the supertrace rule. For example, using Eq. (111) we obtain the character

$$\chi_{\square\square}(U) = \sum_{A,B} (-1)^{g(A)} (-1)^{g(B)} U_{AB}{}^{AB} \quad , \tag{130}$$

$$= 1/2\ (\mathrm{Str}\ U)^2 + 1/2\ \mathrm{Str}(U^2) \quad ,$$

which is obviously invariant under the similarity transformations. Applying the same method we obtain for the adjoint representation

$$\chi_{\square\square}(U) = (\mathrm{Str}\ U^{+})\ (\mathrm{Str}\ U) - 1 \quad . \tag{131}$$

As an exercise, I leave it to the reader to calculate with the methods described above the following quantities:

(a) The matrix representations and

(b) the characters,

corresponding to the following supertableaux, and obtain the given answers. The procedure that yields these results can be extended, of course, to any supertableau.

$$\chi_{\square} = \mathrm{Str}\ U \quad ,$$

$$\chi_{\square\square} = 1/2\ (\mathrm{Str}\ U)^2 + 1/2\ \mathrm{Str}(U^2) \quad ,$$

$$\chi_{\begin{smallmatrix}\square\\\square\end{smallmatrix}} = 1/2\ (\mathrm{Str}\ U)^2 - 1/2\ \mathrm{Str}(U^2) \quad ,$$

$$\chi_{\square\square\square} = 1/6\ (\mathrm{Str}\ U)^3 + 1/2\ \mathrm{Str}\ U\ \mathrm{Str}(U^2) + 1/3\ \mathrm{Str}(U^3) \quad ,$$

$$\chi_{\begin{smallmatrix}\square\square\\\square\end{smallmatrix}} = 1/3\ (\mathrm{Str}\ U)^3 - 1/3\ \mathrm{Str}(U^3) \quad , \tag{132}$$

$$\chi_{\begin{smallmatrix}\square\\\square\\\square\end{smallmatrix}} = 1/3\ (\mathrm{Str} U)^3 - 1/2\ \mathrm{Str}\ U\ \mathrm{Str}(U^2) + 1/3\ \mathrm{Str}(U^3) \quad ,$$

$$\chi_{\square\square\square\square} = 1/24\ (\mathrm{Str}\ U)^4 + 1/4\ (\mathrm{Str}\ U)^2\ \mathrm{Str}(U^2) + 1/3\ \mathrm{Str}\ U\ \mathrm{Str}(U^3) + 1/8\ \left(\mathrm{Str}(U^2)\right)^2 + 1/4\ \mathrm{Str}(U^4) \quad ,$$

$$\chi_{\begin{smallmatrix}\square\\\square\\\square\\\square\end{smallmatrix}} = 1/24\ (\mathrm{Str}\ U)^4 - 1/4\ (\mathrm{Str}\ U)^2\ \mathrm{Str}(U^2) + 1/3\ \mathrm{Str}\ U\ \mathrm{Str}(U^3) + 1/8\ (\mathrm{Str}\ U^2)^2 - 1/4\ \mathrm{Str}(U^4) \quad ,$$

$$\chi_{\begin{smallmatrix}\square\square\square\\\square\end{smallmatrix}} = 1/8\ (\mathrm{Str}\ U)^4 + 1/4\ (\mathrm{Str}\ U)^2\ \mathrm{Str}(u^2) - 1/8\ (\mathrm{Str}\ U^2)^2 - 1/4\ \mathrm{Str}(U^4) \quad ,$$

χ  $= 1/8\ (\mathrm{Str}\ U)^4 - 1/4\ (\mathrm{Str}\ U)^2\ \mathrm{Str}(U^2) - 1/8\ (\mathrm{Str}\ U^2)^2 + 1/4\ \mathrm{Str}(U^4)$,

χ 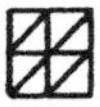 $= 1/12\ (\mathrm{Str}\ U)^4 - 1/3\ \mathrm{Str}\ U\ \mathrm{Str}(U^3) + 1/4\ (\mathrm{Str}\ U^2)^2$,

χ  $= \mathrm{Str}\ U^+$,

χ  $= \mathrm{Str}\ U^+\ \mathrm{Str}\ U - 1$,

χ  $= 1/2\ \mathrm{Str}\ U^+[(\mathrm{Str}\ U)^2 + \mathrm{Str}(U^2)] - \mathrm{Str}\ U$,

χ 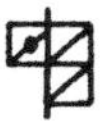 $= 1/2\ \mathrm{Str}\ U^+[(\mathrm{Str}\ U)^2 - \mathrm{Str}(U^2)] - \mathrm{Str}\ U$, (133)

χ 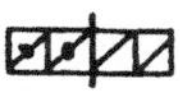 $= 1/4\ [(\mathrm{Str}\ U^+)^2 + \mathrm{Str}(U^{+2})]\ [(\mathrm{Str}\ U)^2 + \mathrm{Str}(U^2)] - \mathrm{Str}\ U^+\ \mathrm{Str}\ U$,

χ 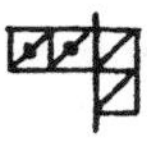 $= 1/4\ [(\mathrm{Str}\ U^+)^2 + \mathrm{Str}(U^{+2})]\ [(\mathrm{Str}\ U)^2 - \mathrm{Str}(U^2)] - \mathrm{Str}\ U^+\ \mathrm{Str}\ U + 1$,

χ 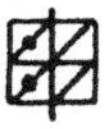 $= 1/4\ ](\mathrm{Str}\ U^+)^2 - \mathrm{Str}(U^{+2})][(\mathrm{Str}\ U)^2 - \mathrm{Str}(U^2)] - \mathrm{Str}\ U^+\ \mathrm{Str}\ U$.

Note that if the same calculations are repeated for SU(N) the answers will formally look the same except for the replacement of supertrace by trace. From this observation, you learn that if you can write the SU(N) answer in terms of $\mathrm{tr}(U^n)$ for some Young tableau you only need to change formally tr to Str to obtain the character for the analogous supertableau!!!

$$SU(N) \dashrightarrow SU(N/M) ,$$
$$\mathrm{tr}\ U^n \dashrightarrow \mathrm{Str}\ U^n . \qquad (134)$$

There is a systematic way of writing down the answer for any given supertableau. First, from the above examples in Eq. (132), note that for a supertableau with a single row with n boxes you can write

χ 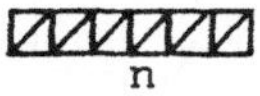 $= 1/n!\ (\mathrm{Str}\ U)^n + 1/(n-2)!(\mathrm{Str}\ U)^{n-2}\ \mathrm{Str}(U^2)/2 + .. + \mathrm{Str}(U^n)/n$

$$= \sum_{k_1 k_2 \ldots k_n} \frac{\delta(n-k_1-2k_2 \ldots nk_n)}{k_1!\, k_2! \ldots k_n!} \left[(\mathrm{Str}U)^{k_1} (\mathrm{Str}(U)^2/2)^{k_2} \ldots (\mathrm{Str}(U^n)/n)^{k_n}\right] . \tag{135}$$

This sum can be rewritten as an integral, and it depends only on U and the number of superboxes n. We will denote $\chi_n \equiv \chi$ 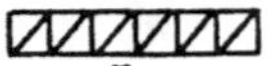 n

n 

$$\chi_n(U) = \int_0^{2\pi} \frac{d\phi}{2\pi} \frac{e^{-in\phi}}{\mathrm{Sdet}\,(1-e^{i\phi}U)} \; ; \; n \geq 0 , \tag{136}$$

where for $n = 0$ $\chi_0 = 1$ and for negative integers $\chi_n = 0$ $(n < 0)$ automatically.

If all the boxes in a single row are replaced by dotted boxes (upper indices), then every U is replaced by U^+ in the transformation property of the tensor. Thus, the character of the corresponding group element will have the same form as $\chi_n(U)$, except for U replaced by U^+.

We will denote this expression by $\dot{\chi}_{-n}(U)$ as a convenient notation

$$\dot{\chi}_{-n}(U) \equiv \chi_n(U^+) ,$$

$$\dot{\chi}_{-n}(U) = \int_0^{2\pi} \frac{d\phi}{2\pi} \frac{e^{in\phi}}{\mathrm{Sdet}\,(1-e^{i\phi}U^+)} . \tag{137}$$

Returning to the examples above in Eq. (132), we observe that for two and three rows of undotted boxes the answer is reproduced correctly by the following determinants

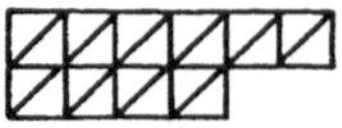 n_1 n_2

$$\chi_{n_1,n_2}(U) = \begin{vmatrix} \chi_{n_1} & \chi_{n_2-1} \\ \chi_{n_1+1} & \chi_{n_2} \end{vmatrix} ,$$

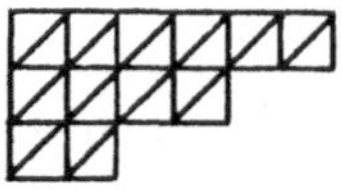 n_1 n_2 n_3

$$\chi_{n_1,n_2,n_3}(U) = \begin{vmatrix} \chi_{n_1} & \chi_{n_2-1} & \chi_{n_3-2} \\ \chi_{n_1+1} & \chi_{n_2} & \chi_{n_3-1} \\ \chi_{n_1+2} & \chi_{n_2+1} & \chi_{n_3} \end{vmatrix} . \tag{138}$$

It is clear from the systematics of these expressions that for r rows we need to evaluate an r x r determinant of a matrix whose (i,j) elements are shown

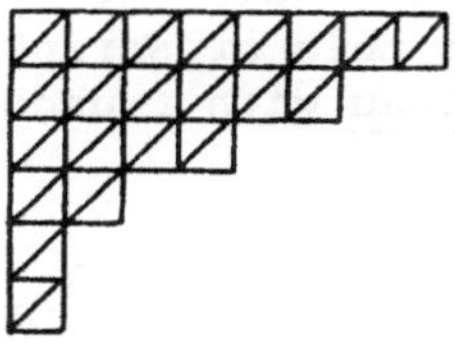

n_1, n_2, ..., n_r

$$\chi_{(n_1,\, n_2\, \ldots\, n_r)} = \det\left(\chi_{n_j + i - j}\right) . \tag{139}$$

This form, which was derived for supergroups[14] on the basis of the above observations, is also a familiar form[20] for ordinary SU(N) and is appropriate for the analogy drawn in Eq. (134): For SU(N) indeed the correct character is reproduced by Eq. (139) and hence for SU(N/M) as well.

Now we also consider the mixed dotted-undotted supertableaux. We observe from the examples above that the correct answers are reproduced by the following determinants

m_1 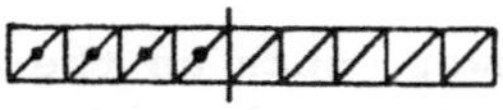 n_1

$$\chi_{(-m_1;n_1)} = \begin{vmatrix} \dot{\chi}_{-m_1} & \chi_{n_1+1} \\ \dot{\chi}_{-m_1+1} & \chi_{n_1} \end{vmatrix} ,$$

m_1 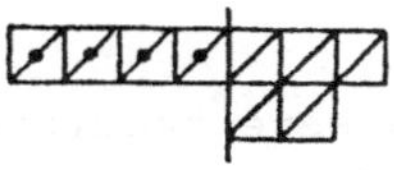 m_1, m_2

$$\chi_{(-m_1;n_1,n_2)} = \begin{vmatrix} \dot{\chi}_{-m_1} & \chi_{n_1-1} & \chi_{n_2-2} \\ \dot{\chi}_{-m_1+1} & \chi_{n_1} & \chi_{n_2-1} \\ \dot{\chi}_{-m_1+2} & \chi_{n_1+1} & \chi_{n_2} \end{vmatrix} , \tag{140}$$

m_1, m_2 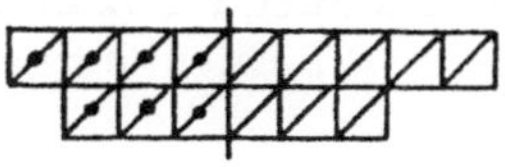 n_2, n_2

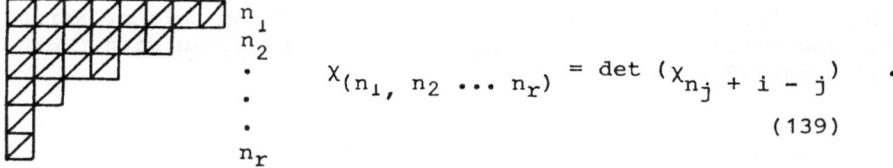

$$\chi_{(-m_2,-m_1,n_1,n_2)} = \begin{vmatrix} \dot{\chi}_{-m_2} & \dot{\chi}_{-m_1-1} & \chi_{n_1-2} & \chi_{n_2-3} \\ \dot{\chi}_{-m_2+1} & \dot{\chi}_{-m_1} & \chi_{n_1-1} & \chi_{n_2-2} \\ \dot{\chi}_{-m_2+2} & \dot{\chi}_{-m_1+1} & \chi_{n_1} & \chi_{n_2-1} \\ \dot{\chi}_{-m_2+3} & \dot{\chi}_{-m_1+2} & \chi_{n_1+1} & \chi_{n_2} \end{vmatrix} . \tag{141}$$

The systematics of the mixed tableaux are seen to be the same as the undotted tableaux: On the diagonal we write the character of a row starting with the shortest dotted row and ending with the shortest undotted row. The rest of the determinant is filled by increasing the index downward or decreasing it upward. Note again the definition of $\dot{\chi}_{-m}(U) = \chi_m(U^+)$ as in Eq. (137). Using this prescription we can now compute the character for any supertableau with r undotted and $\dot{r}$ dotted rows.

$$\chi_{(-m_{\dot{r}},\ldots,-m_1;\ n_1,\ldots n_r)} = \begin{vmatrix} \dot{\chi}_{-m_{\dot{r}}} & & & & & \chi_{n_r-r-r+1} \\ & \ddots & & & \text{decrease} \ \Uparrow & \\ \Downarrow & & \dot{\chi}_{-m_1} & & & \\ & & & \chi_{n_1} & & \\ \text{increase} & & & & \ddots & \\ \dot{\chi}_{-m_{\dot{r}}+\dot{r}+r-1} & & & & & \chi_{n_r} \end{vmatrix} . \qquad (142)$$

This formula gives a very explicit answer, such as the examples of Eqs. (132-133), which is very useful in practical calculations. This new expression for characters for mixed tableaux can also be applied in SU(N) because of the analogy of Eq. (134). It appears that this is a new form[14] not only for SU(N/M) but also for SU(N). Its validity is checked by comparing to alternative forms of the characters and is seen to work, as expected.

In the above expressions the supertableaux were specified by the number of boxes in each row and the characters were calculated accordingly. A similar procedure works if the labeling is done by the number of boxes in each column. Returning to the example in Eq. (132) we observe that for a single column of undotted boxes we can write

χ  n

$$= (-1)^n \sum_{k_1,\ldots,k_n} \frac{\delta(n-k_1-2k_2\ldots-nk_n)}{k_1!\ k_2!\ \ldots\ k_n!} \left[(-\mathrm{Str}U)^{k_1}\ (-\mathrm{Str}(U^2)/2)^{k_2}\ \ldots\ (-\mathrm{Str}(U^n)/n)^{k_n}\right] . \qquad (143)$$

This result can also be obtained[14] by evaluating an n x n determinant according to the formula in Eq. (139). We will denote this expression by $A_n(U)$ and note that it can be written as an integral, in analogy to Eq. (136)

$$An(U) \equiv \chi_{\;\;n}(U) \quad ,$$

$$An(U) = (-1)^n \int_0^{2\pi} \frac{d\phi}{2\pi}\; e^{-in\phi}\; \mathrm{Sdet}(1-e^{i\phi}\,U) \quad . \qquad (144)$$

As in the case of rows, $A_0 = 1$ and $A_n = 0$ if $n \;\; 0$, follows automatically from this equation. It is clear that it would be very cumbersome to use an n x n determinant instead of this simple form. Similar simplifications would occur if we develop a formula appropriate to tableaux with long columns. Then following the steps above and substituting columns instead of rows we verify that the explicit examples in Eq. (132) are correctly reproduced by

$k_1 k_2$

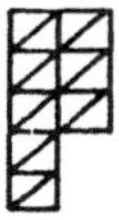

$$A_{k_1,k_2}(U) = \begin{vmatrix} A_{n_1} & A_{n_2-1} \\ A_{n_1+1} & A_{n_2} \end{vmatrix} \quad , \text{ etc.} \qquad (145)$$

The systematics with columns are the same as with the rows, and the character for the general mixed supertableau of Fig. (1) can be evaluated according to the number of boxes in each <u>column</u> in Fig. (1),

$$A_{(-\ell_q,\ldots,-\ell_1;k_1,\ldots,k_p)} = \chi_{(-\dot{m}_r,\ldots,-m_1;\; n_1,\ldots n_r)} \quad ,$$

$$A_{(-\ell_q,\ldots,-\ell_1;k_1,\ldots,k_p)} = \begin{vmatrix} \dot{A}_{-\ell_q} & & & \text{decrease} \\ & \ddots \;\Uparrow & & \\ \Downarrow & & \dot{A}_{-\ell_1} & \\ & & A_{k_1} \;\Uparrow & \\ & & & \ddots \\ \text{increase} & & \Downarrow & A_{k_p} \end{vmatrix} ,\qquad (146)$$

where $\dot{A}_{-m}(U) = A_m(U^+)$ is a definition. The answer is, of course, the same when the character is evaluated via the rows or columns of the supertableau. As exercises the reader can return to Eqs. (132-133).

13. Dimensions

In a given representation we are often interested in the number of states in the vector space. For supergroups, we will need the numbers of bosons and fermions separately. Let us define

$$\begin{aligned}
&\text{number of bosons} && \equiv B \ , \\
&\text{number of fermions} && \equiv F \ , \\
&\text{difference} && \equiv d = B - F \ , \\
&\text{sum} && \equiv S = B + F \ .
\end{aligned} \tag{147}$$

We will be able to compute d and S rather simply. In some simple cases above we have already counted these states. For example, in the fundamental representation

$$d_{\square} = N - M \ , \qquad S_{\square} = N + M \ . \tag{148}$$

Also, from Eqs. (106) and (107) we have

$$\begin{aligned}
d_{\square\square} &= 1/2\ (N - M)(N - M + 1) \ , \\
S_{\square\square} &= 1/2\ \left[(N + M)^2 + N - M\right]
\end{aligned} \tag{149}$$

and

$$\begin{aligned}
d_{\begin{smallmatrix}\square\\\square\end{smallmatrix}} &= 1/2\ (N - M)(N - M - 1) \ , \\
S_{\begin{smallmatrix}\square\\\square\end{smallmatrix}} &= 1/2\ \left[(N + M)^2 - (N - M)\right] \ .
\end{aligned} \tag{150}$$

Rather than counting the states as in Eqs. (106,107) we can do our computation much more efficiently by realizing that d and S follow from the evaluation of the character for the identity matrix $U = 1$ and for the matrix $U = J = \left(\begin{array}{c|c} 1 & \\ \hline & -1 \end{array}\right)$ respectively. For example, for the fundamental representation

$$\begin{aligned}
d_{\square} &= \mathrm{Str}\ 1 = N - M \ , \\
S_{\square} &= \mathrm{Str}\ J = N + M \ .
\end{aligned} \tag{151}$$

Similarly, using the character formulas in Eq. (130), we have.

$$d \; \text{(tableau)} = 1/2 \left[(\mathrm{Str}\ 1)^2 + \mathrm{Str}(1^2)\right] = 1/2 \left[(N - M)^2 + (N - M)\right] ,$$

(152)

$$S \; \text{(tableau)} = 1/2 \left[(\mathrm{Str}\ J)^2 + \mathrm{Str}(J^2)\right] = 1/2 \left[(N + M)^2 + (N - M)\right] ,$$

in agreement with Eq. (149). Similarly, for the adjoint representation

$$d \; \text{(tableau)} = (N - M)^2 - 1 = (\mathrm{Str}\ 1)^2 - 1 ,$$

(153)

$$S \; \text{(tableau)} = (N + M)^2 - 1 = (\mathrm{Str}\ J)^2 - 1 ,$$

which coincide with the counting of bose + fermi parameters in the super algebra given by H in Eq. (1), as expected. The same can be checked explicitly for all the representations of Eqs. (132,133). For an arbitrary representation R we can now make the statement

$$d_R = \chi_R(1) \quad , \qquad S_R = \chi_R(J) \quad . \tag{154}$$

These can be computed from the character formulas above provided we first compute $\chi_n(1)$ and $\chi_n(J)$ for one row (and/or for one column).

Thus, in general we need to evaluate d and S for the single column and single row supertableaux and simply substitute the result in the general character formulas of Eqs. (142) or (146). We define the dimensions for one row and one column as

n
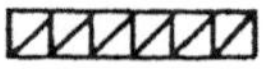
$\dashrightarrow \; d_n \; , \; Sn \quad ,$

(155)

n

$\dashrightarrow \; \tilde{d}_n \; , \quad \tilde{S}_n \quad .$

In an obvious notation that follows Eq. (137) we obtain for a dotted row or column

n 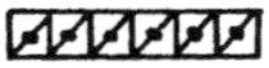 $\dashrightarrow \dot{d}_{-n} \;,\; \dot{s}_{-n} \;,$ with $\begin{cases} \dot{d}_{-n} = d_n \\ \dot{s}_{-n} = s_n \end{cases}$ (156)

n  $\dashrightarrow \tilde{\dot{d}}_{-n} \;,\; \tilde{\dot{s}}_{-n} \;,$ with $\begin{cases} \tilde{\dot{d}}_{-n} = d_n \\ \tilde{\dot{s}}_{-n} = s_n \end{cases}$.

We evaluate them directly from the integral representations in Eqs. (136) for one row

$$d_n = \int_0^{2\pi} \frac{d\phi}{2\pi}\, e^{-in\phi}\, (1-e^{i\phi})^{M-N} = \frac{1}{n!}(N-M)(N-M+1)\ \ldots\ (N-M+n-1) \quad , \tag{157}$$

$$s_n = \int_0^{2\pi} \frac{d\phi}{2\pi}\, e^{in\phi}\, \frac{(1+e^{i\phi})^M}{(1-e^{i\phi})^N} = \sum_k \binom{N+n-k-1}{N-1}\binom{M}{k} \quad ,$$

where $\binom{b}{a} = \frac{b!}{a!(b-a)!}$ is the binomial coefficient. Similarly $\tilde{d}_n$, $\tilde{s}_n$ for one column are given via Eq. (144)

$$\tilde{d}_n = (-1)^n \int_0^{2\pi} \frac{d\phi}{2\pi}\, e^{-in\phi}\, (1-e^{i\phi})^{N-M} = \frac{1}{n!}(N-M)(N-M-1)\ldots(N-M-n+1) \tag{158}$$

$$\tilde{s}_n = (-1)^n \int_0^{2\pi} \frac{d\phi}{2\pi}\, e^{in\phi}\, \frac{(1-e^{i\phi})^N}{(1+e^{i\phi})^M} = \sum_k \binom{N}{k}\binom{M+n-k-1}{M-1} \quad .$$

It is striking that the formulas for $\tilde{d}_n$ and d_n coincide with those of the dimension formulas for the group SU(N-M) for the one row and column <u>ordinary</u> Young tableaux!!

$d_n\left(SU(N/M)\right)$ = dimension for n 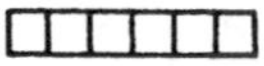 of SU(N-M) , (159)

$$\tilde{d}_n\left(SU(N/M)\right) = \text{dimension for } \begin{matrix}\Box\\\Box\\\Box\end{matrix}\; n \text{ of } SU(N-M) \quad .$$

While the formulas for S_n, $\tilde{S}_n$ coincide with the sum of dimensions of the SU(N) x SU(M) tableaux

$$S_n\left(SU(N/M)\right) = \text{dimension for } \sum_k \Big(\overbrace{\Box\Box\Box\Box\Box\Box}^{n-k} \;,\; \begin{matrix}\Box\\\Box\\\Box\end{matrix}\; k \Big) \quad \text{for } SU(N) \times SU(M) \quad ,$$

$$\tilde{S}_n\left(SU(N/M)\right) = \text{dimension for } \sum_k \Big(\begin{matrix}\Box\\\Box\\\Box\\\Box\end{matrix}\; k,\; \overbrace{\Box\Box\Box\Box\Box}^{n-k} \Big) \quad \text{for } SU(N) \times SU(M) \quad . \qquad (160)$$

The general dimension formulas for the supertableau of Fig. (1) can now be given via Eqs. (142) or (146) and (155 - 160), by using columns or rows

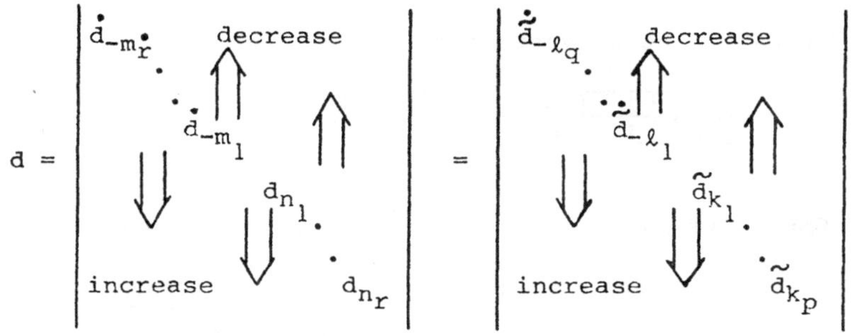

(161)

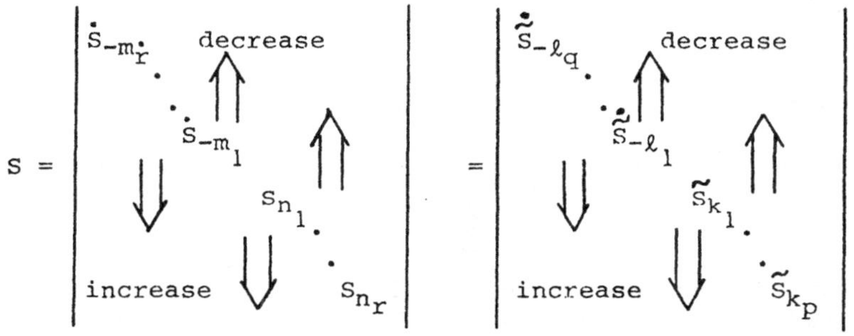

(162)

It is very useful to note that, because of Eq. (159), the number or bosons <u>minus</u> fermions d_R in a supertableau R is numerically identical to the dimension of a representation of the ordinary group SU(N-M) provided the same shape ordinary Young tableau is used. The following few striking examples will show the significance of this observation:

(1) For U(1) the only posible nonvanishing tableaux have n boxes: 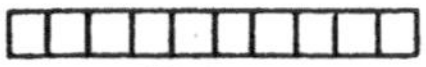 . The dimension of the representation is 1 for any n. Two or more rows will have vanishing dimension. Therefore for SU(N+1/N) we have

n  Bosons - Fermions = 1 $n \geqslant 0$,

n_1 n_2 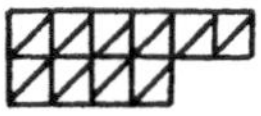 Bosons - Fermions = 0 $n_1 \geqslant n_2 \geqslant 1$,

etc. for more rows.

Thus the number of bosons is equal to the number of fermions for two or more rows. Note that the two or more row supertableaux do not vanish for SU(N+1/N), unlike U(1). The total number of (bosons + fermions) is non-zero as can be checked from Eq. (162).

(2) For SU(2) the nonvanishing tableaux are

SU(2) 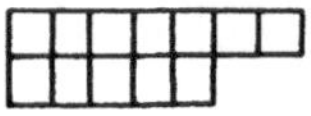 n_1 n_2 dimension = n_1-n_2+1 .

Therefore, for SU(N+2/N) the sample shape supertableau has

Bosons - Fermions = n_1-n_2+1 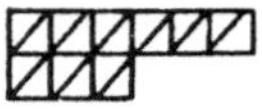 n_1 n_2 .

For three or more rows in the supertableau we have

Bosons - Fermions = 0 ,

while the total number S is nonvanishing.

(3) For SU(5)  is 10 dimensional, therefore for SU(N+5/N)

Bosons - Fermions = 0  .

The examples can go on indefinitely. An important result is that for SU(N/M), if the supertableau has more than (N-M) rows, then the number of bosons is equal to the number of fermions in that representation. These include the "typical" representations that Kac has described in a very different language. The dimension formula for

Bosons + Fermions = S ,

for (d=0) as well non-typical (d≠0) representations is given via Eq. (162). "Typical" representations[19] have a SU(N/M) dimension formula d=0 and $S=2^{NM} \times D$ where D is the dimension of the irreducible SU(N) x SU(M) representation that contains the highest weight of the superrepresentation. Not all of our d=0 representations satisfy this formula for S so that although some such representations are typical, others are "non-typical" even though they contain equal numbers of bosons and fermions (d = 0).

For example for SU(N+1/N) or SU(N/N+1) the adjoint representation ▱ satisfies d ▱ = 0 while $S = (2N+1)^2 - 1$. Furthermore, the adjoint representations of SU(N+1/N) satisfy the following branching to SU(N+1) x SU(M) x U(1)

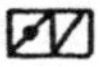

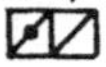

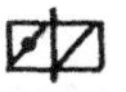

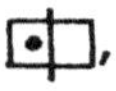

$$▱ = (\square, \boxdot) + (\boxdot, 1) + (1, 1) + (1, \boxdot)_0 + (\boxdot, \square)$$

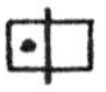

with the first piece ($\square$, $\boxdot$) containing the highest weight. The dimension of this piece is D=(N+1)N. Thus, we may ask for which SU(N+1/N) or SU(N/N+1) is the adjoint representation a typical representation. We must satisfy the requirement for the form of the dimension formula $S=2^{(N+1)N} D$, giving the condition

$$S = (2N+1)^2 - 1 = 2^{(N+1)N} (N+1)N \quad ,$$

which is satisfied for N=1. Thus only for SU(2/1) or SU(1/2) the adjoint representation is a "typical" representation. For all other SU(N+1/N) or SU(N/N+1) the adjoint representation is not "typical", but still contains an equal number of bosons and fermions.

14. Eigenvalues of Casimir Operators

The quantum Casimir operators or order n, $\hat{C}_n$, were constructed in Eqs. (70, 71, 74, 75, 76, 85). By building the states of a representation via the quantum oscillators as in Eqs. (113-122) and further generalizations involving all the oscillators ξ, η, ρ, σ, we may compute the eigenvalues. This works very efficiently for representations of the type ▱▱▱▱ as in Eq. (122), since, as we argued following Eq. (77), in this case there really is a single independent invariant, which is $\hat{n} = \xi^+ \cdot \xi$. For example, applying the quadratic Casimir operator $\hat{C}_2$ in Eq. (74) to the state in Eq. (122) we obtain immediately the eigenvalue for the supergroup SU(N/M)

$$C_2(\underbrace{▱▱▱▱}_{n}) = \frac{N-M-1}{2(N-M)} n(N-M+n) \; ; \quad N \neq M \quad . \tag{163}$$

Note that for N=M, C_2 could not be defined, but the invariant operator n and its eigenvalue exists. The computation of $C_2(R)$ is not so easy if a few copies of oscillators are involved as can be seen from

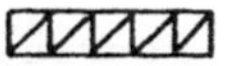

Eq. (71).

A different general approach was described in Ref. [14], which uses the characters in Eqs. (142, 146) as the starting point: Recall that the group parameters are described by the matrix H in the fundamental representation, as in Eqs. (1, 2, 35)

$$H_A{}^B = (\lambda_I/2)_A{}^B \, \omega^I \ , \qquad U = e^{iH} \ , \tag{164}$$

where $(\lambda_I/2)_A{}^B$ represent the generators X_I. In an arbitrary representation R the generator will have a matrix representation $R(X_I)$.

Combining generators with parameters, we can then write the representation of the matrix H as

$$R_p{}^q(H) = R_p{}^q\,(X_I)\,\omega^I \ , \tag{165}$$

where the indices p, q run over the bosonic and fermionic states of the representation space. The matrix representation of the generators $\underline{X}_A{}^B$ or $\underline{G}_A{}^B$ defined in Eqs. (66) or (82) can now be obtained by differentiating with respect to the parameters $H_A{}^B$ as in Eq. (1, 2)

$$R_p{}^q\,(\underline{X}_A{}^B) = \frac{\partial}{\partial H_B{}^A}\, R_p{}^q\,(H) \ . \tag{166}$$

Writing the n'th order Casimir operator as Eq. (76)

$$\hat{C}_n = \mathrm{Str}(\underline{X}^n) = \underline{X}_{A_1}{}^{A_2}\,\underline{X}_{A_2}{}^{A_3} \cdots \underline{X}_{A_n}{}^{A_1}\,(-1)^{g(A_1)} \ , \tag{167}$$

we see that its matrix representation will be given by

$$R\,{}_p{}^q\,(\hat{C}_n) = (-1)^{g(A_1)}\frac{\partial}{\partial H_{A_1}{}^{A_n}} \cdots \frac{\partial}{\partial H_{A_3}{}^{A_2}}\,\frac{\partial}{\partial H_{A_2}{}^{A_1}}\quad R\,{}_p{}^q\,(H^n) \ . \tag{168}$$

The matrix $R_p{}^q(\hat{C}_n)$ is found proportional to the identity for $d(R) \neq 0$ representations. While it probably has non-diagonal parts for $d_R=0$ representations, it can be argued that Shur's lemma works only in representation in which the number of bosons is different from the number of fermions ($d_R \neq 0$). Then we may write

$$R_p{}^q\,(\hat{C}_n) = C_n(R)\,\delta_p{}^q \ , \qquad d_R \neq 0 \ . \tag{169}$$

Our aim is to obtain the number $C_n(R)$. For $d(R) = 0$ representations, since there could be a non-diagonal piece our method will not work.

Now we describe the method of computation and give the answer.

By taking the supertrace of Eq. (169), we obtain

$$\mathrm{Str}\ R(\hat{C}_n) = C_n(R)\ d_R \quad ,$$

or

$$C_n(R) = \mathrm{Str}\ R(\hat{C}_n)\ /\ d_R \ ; \quad d_R \neq 0 \quad , \tag{170}$$

so that we only need to compute Str $R(\hat{C}_n)$ which is given, from Eq. (168), by

$$\mathrm{Str}\ R(\hat{C}_n) = \left[\mathrm{Str}(\partial/\partial H)^n \left[\mathrm{Str}\ R(H^n)\right]\right. \quad , \tag{171}$$

where we wrote the differential operators symbolically. To calculate this quantity we return to the character which we have already computed. The aim is to bypass the need of constructing the matrix element $R_p{}^q\ (H)$ explicitly, and use only the character which is much simpler.

The characters of Eqs. (142, 146), which depend explicitly on the parameters $H_A{}^B$, can also be expressed in a different form by computing the supertrace of the representation of supergroup elements in the notation of Eq. (165):

$$\text{group:} \qquad R_p{}^q\ (U) = \left[e^{iR(H)}\right]_p{}^q \quad ,$$

$$\text{character:} \quad \chi_R\ (U) = \mathrm{Str}\left[e^{iR(H)}\right] \quad . \tag{172}$$

Then, expanding the exponential in powers of R(H) it is clear that Eq. (171) follows from

$$\mathrm{Str}\ R(\hat{C}_n) = n!/(i)^n \left\{\left[\mathrm{Str}\ (\partial/\partial H)^n\right] \chi_R(U)\right\}_{H=0} \quad , \tag{173}$$

where $H_A{}^B = 0$ after differentiating n times. But now the desired quantity has been expressed in terms of $\chi_R(U)$ and we no longer need R(H).

Simply, we can substitute Eq. (142) or (146) for $\chi_R(U)$ which is quite explicit in terms of H. Thus, we use our explicit character formulas, substitute $U = e^{iH}$ and expand in powers of H. (As an exercise you can do this with a few examples provided by Eqs. (132-133)). This expansion may be written in two forms, the first following from Eq. (172) and the second from (142, 146)

$$\chi_R(e^{iH}) \nearrow \ \mathrm{Str}\ R(1) + i^2/2!\ \mathrm{Str}\left[R(H)\right]^2 + i^3/3!\ \mathrm{Str}\left[R(H)\right]^3 + i^4/4!\ \mathrm{Str}\left[R(H)\right]^4 + \dots$$

$$\chi_R(e^{iH}) \searrow \ d_R + q_R/2!\ \mathrm{Str}(iH)^2 + k_R/3!\ \mathrm{Str}(iH)^3 + \{v_R/4!\ \mathrm{Str}(iH)^4$$

$$+ w_R/2!\ (1/2\ \mathrm{Str}(iH)^2)^2\} + \ldots \tag{174}$$

where we have taken into account that Str H = 0. Thus, in order to compute Eq. (173) we will need to first find the coefficients q(R), k(R), v(R), W(R) etc. via Eqs. (142, 146) which will be given below.

From this form we show how to compute, say, the quadratic Casimir. From eqs. (170, 173, 174) we have for a representation with $d_R \neq 0$

$$C_2(R) = \frac{1}{d_R}\, q_R \left[\mathrm{Str}(\partial/\partial H)^2\right]\left[\mathrm{Str}\ H^2\right] \quad . \tag{175}$$

The last factor, as defined through Eqs. (170-171), is simply the Casimir in the fundamental representation $C_2\ (\boxslash)$ times $d_{\boxslash}$, which we can compute easily, as in Eq. (49). Therefore

$$C_2(R) = \frac{q_R}{d_R}\ C_2\ (\boxslash)\ d(\boxslash) \quad . \tag{176}$$

Similarly, we obtain the cubic Casimir as

$$C_3(R) = \frac{k_R}{d_R}\ C_3(\boxslash)\ d(\boxslash) \tag{177}$$

and so on for the higher Casimirs.

Now, we turn to the computation of q(R), k(R) etc. First we need to compute the expansion of χ_n for one row and/or A_n for one column, and then substitute them in the general character formulas (142, 146). For one row or column in Eqs. (136, 144) we substitute $U = e^{iH}$ and expand in powers of H. By doing some simple integrals we get[14] :

$$\begin{aligned} q_{\underbrace{\boxslash\boxslash\boxslash\boxslash\boxslash\boxslash\boxslash\boxslash}_{n}} &\equiv q_n = \frac{n(N - M + n)}{(N-M)(N-M+1)}\ d_n \quad , \\ k_{\underbrace{\boxslash\boxslash\boxslash\boxslash\boxslash\boxslash\boxslash\boxslash}_{n}} &\equiv k_n = \frac{n(N-M+n)(N-M+2n)}{(N-M)(N-M+1)(N-M+2)}\ d_n \quad , \end{aligned} \tag{178}$$

$$q_{\left.\begin{smallmatrix}\boxslash\\\boxslash\\\boxslash\\\boxslash\\\boxslash\\\boxslash\\\boxslash\end{smallmatrix}\right\} n} \equiv \tilde{q}_n = \frac{n(N-M-n)}{(N-M)(N-M-1)}\ \tilde{d}_n \quad , \tag{179}$$

k  n $\equiv \tilde{k}_n = \dfrac{n(N-M-n)(N-M-2n)}{(N-M)(N-M-1)(N-M-2)}\;\tilde{d}_n$

and so on for v_n, w_n, etc., where d_n, $\tilde{d}_n$ are already given in Eqs. (157, 158). Thus, for 1 row or column Eq. (174) takes the form

$$\chi_n(e^{iH}) = d_n + q_n \frac{\mathrm{Str}(iH)^2}{2!} + k_n \frac{\mathrm{Str}(iH)^3}{3!} + \ldots \quad ,$$

(180)

$$A_n(e^{iH}) = d_n + q_n \frac{\mathrm{Str}(iH)^2}{2!} + k_n \frac{\mathrm{Str}(iH)^3}{3!} + \ldots \quad .$$

Substituting these in the general forms Eqs. (142, 146), reexpanding in powers of H and comparing to Eq. (174) we obtain the general q(R), k(R) etc. in the form of a sum of determinants. Symbolically this looks like

$$X(R) = \begin{vmatrix} x & d & d & \cdots \\ x & d & d & \cdots \\ x & d & d & \cdots \\ \cdot & \cdot & \cdot & \\ \cdot & \cdot & \cdot & \\ \cdot & \cdot & \cdot & \end{vmatrix} + \begin{vmatrix} d & x & d & \cdots \\ d & x & d & \cdots \\ d & x & d & \cdots \\ \cdot & \cdot & \cdot & \\ \cdot & \cdot & \cdot & \\ \cdot & \cdot & \cdot & \end{vmatrix} + \begin{vmatrix} d & d & x & \cdots \\ d & d & x & \cdots \\ d & d & x & \cdots \\ \cdot & \cdot & \cdot & \\ \cdot & \cdot & \cdot & \\ \cdot & \cdot & \cdot & \end{vmatrix}$$

(181)

where X(R) stands for q(R) or k(R) if Eq. (142) is used, or for $\tilde{q}(R)$ or $\tilde{k}(R)$ if Eq. (146) is used. In the case of v(R), w(R) and higher coefficients, the pattern of determinants is more complicated, because the Taylor expansion of the determinant in powers of H is more involved for H^4 and higher powers.

Here we can make an important observation[14 15] that all of the coefficients d(R), q(R), k(R), v(R) etc., for all representations R are coming out as functions of only (N-M) as seen from Eqs. (178-181). Indeed has we calculated all the analogous coefficients d(R), q(R) etc. and Casimir eigenvalues for the ordinary group SU(N-M) for the same shape Young tableau we would get identical numerical expressions as above by using exactly the same formalism step by step. This supplements the previous observation in Eqs. (159, 161) about d(R) = (Bosons - Fermions). Thus, we have the correspondence between SU(N-M) and S(N/M):

$$SU(N-M) \longrightarrow \left[\begin{array}{l} d_R,\ q_R,\ \tilde{q}_R,\ v_R,\ w_R \text{ etc., and} \\ \text{Casimir eigenvalues for } R\ (d_R \neq 0) \end{array} \right] \longrightarrow SU(N/M) \tag{182}$$

Note that the correspondence for the expansion coefficients d(R), q(R), k(R), etc. hold for $d_R = 0$ as well as $d_R \neq 0$ representations, since they occur in the expansion of the character and are independent from the validity of Eq. (169) and (170) for the Casimir operators in which we need to divide by d_R.

From the above statements we now observe[6 14 15] that if we have an expression for the Casimir eigenvalues for the group SU(N), written as a function of N, we could simply substitute $N \to N - M$ in order to obtain the Casimir eigenvalues for SU(N/M). Indeed, making use of this trick, all Casimir eigenvalues for up to the (N-M)th order Casimir have been computed[15] by simply translating $N \to N-M$ in the expressions available in the literature on SU(N). The answer is expressed in the form of a generating functional that yields the eigenvalues of the p'th order SU(N/M) Casimir operator for $p \leq |N-M|$

$$C_p = \mathrm{Str}(\underline{X}^p) \qquad p \leq |N - M|$$

supertableau

$$R = (n_1, n_2, n_3, \ldots) \quad ,$$

then the generating functional is[21]

$$\sum_{p=0}^{\infty} C_p(R)\, Z^p = (N-M)\, e^{-f(Z)} + (1-e^{-f(Z)})/Z$$

where

$$f(Z) = \sum_{k=2}^{\infty} \sum_{j=1}^{k-1} \frac{Z^k}{k} \binom{k}{j} S_J \quad ,$$

$$S_j = \sum_{i=1}^{N-M} \sum_{\ell=0}^{j-1} \binom{j}{\ell} \left(n_i - \frac{\sum_r n_r}{N-M} \right)^{j-\ell} (N-M-i)^{\ell} \quad .$$

In comparison to the determinental forms Eq. (181), the quadratic Casimir eigenvalues computed in this way are simpler, and given by[15]

$$C_2(R) = 1/2 \sum_{i=1}^{\infty} \left[n_i^2 - 2in_i + (N-M+1)\, n_i \right] - \frac{1}{2(N-M)} \left(\sum_{r=1}^{\infty} n_r \right)^2 \quad .$$

Note that, unlike SU(N), the sum extends to infinity. We have generalized this formula to the general supertableau in Fig. (1) containing dotted boxes as well. Thus, for the quadratic Casimir we obtain (for $d_R \neq 0$) the new result

$$C_2(R) = 1/2 \sum_{i=1}^{\infty} \{n_i^2+m_i^2+(n_i+m_i)(N-M+1-2i)\} - \frac{1}{2(N-M)} [\sum_{r=1}^{\infty} (n_r-m_r)]^2 .$$

15. Branching Rules SU(N/M) ⟶ SU(N) ⊗ SU(M) ⊗ U(1)

This branching rule of the superrepresentations is of primary importance in physical applications[4,6]. This decomposition is also the key for establishing the relation between supertableaux and Kac-Dynkin diagrams. The irreducibility (or reducibility but indecomposability in some cases) properties of our representations, which was discussed to a limited extent in our previous work, becomes evident after this connection.

As in the previous sections, we continue to make progress by further exploiting the relationship between SU(N/M) and SU(N+M). Here we will compare the branching rules for

$$SU(N+M) \supset SU(N) \otimes SU(M) \otimes U(1)$$

and

$$SU(N/M) \supset SU(N) \otimes SU(M) \otimes U(1) \quad . \tag{183}$$

In the N+M dimensional fundamental representation, the U(1) generator is a traceless matrix for SU(N+M) and a supertraceless one for SU(N/M). Up to an overall constant, it is given as:

$$U(1) : \left[\begin{array}{c|c} 1/N & 0 \\ \hline 0 & -1/M \end{array}\right] \text{ for } SU(N+M) \quad ,$$

$$U(1) : \left[\begin{array}{c|c} 1/N & 0 \\ \hline 0 & 1/M \end{array}\right] \text{ for } SU(N/M) \quad . \tag{184}$$

In the following we will work our way up by starting with a few simple examples and eventually arrive at some general observations that hold for any representation.

The fundamental representation $\phi_A \sim \square$ (or ⧄) can be split into the direct sum $\phi_A = \phi_a \oplus \psi_\alpha$. Here the N dimensional piece ϕ_a $a=1,2,\ldots,N$ transforms like the fundamental representation of SU(N), is singlet under SU(M), and carries the U(1) charge 1/N. We denote this part by $\phi_a \sim (\square, 1)_{1/N}$. Similarly the M dimensional piece ψ_α $\alpha=1,2,\ldots,M$ belongs to the fundamental representation of SU(M), is a singlet under SU(N) and carries the U(1) charge (-1/M) for SU(N+M) and 1/M for SU(N/M). We denote it by $\psi_\alpha \sim (1, \square)_{-1/M}$ or $(1, \square)_{1/M}$ respectively. For SU(N+M) ϕ_a and ψ_α are both bosons (or both fermions). For SU(N/M) one of them is a boson and the other is a fermion. In representations of class I we chose ϕ = boson and ψ = fermion. It is sufficient to restrict ourselves to only class I representations since all other representations (class II and mixed cases) can be obtained from those of pure class I representations by simply switching bosons and fermions in the final basis without changing the matrix representation of the group element.

From the above explanation, the branching eq. (183) for the fundamental representation $\phi_A = \phi_a \oplus \psi_\alpha$ can be expressed in terms of tableaux as

$$\begin{aligned} \square &= (\square, 1)_{1/N} \oplus (1, \square)_{-1/M} \text{ for SU(N+M)} \quad , \\ \boxslash &= (\square, 1)_{1/N} \oplus (1, \square)_{1/M} \text{ for SU(N/M)} \quad . \end{aligned} \tag{185}$$

Next consider the completely symmetric (completely supersymmetric) tensor with n indices $\phi(A_1A_2\dots A_n)$. By specializing each index $A_i = a_i \oplus \alpha_i$, where $a_i = 1,2,\dots,N$ and $\alpha_i = 1,2,\dots,M$, we can find the various SU(N) x SU(M) x U(1) components of this tensor as

$$\begin{aligned} \phi(A_1A_2\dots A_n) &= \phi(a_1a_2\dots a_n) \oplus \phi(a_1a_2\dots a_{n-1})(\alpha_n) \oplus \cdots \oplus \quad , \\ &\oplus \phi(a_1a_2)(\alpha_3\alpha_4\dots\alpha_n) \oplus \phi(a_1)(\alpha_2\alpha_3\dots\alpha_n) \oplus \phi(\alpha_1\alpha_2\dots\alpha_n) \quad . \end{aligned} \tag{186}$$

The U(1) charges can be computed by assigning 1/N to each a_i and -1/M(+1/M) to each α_i. For SU(N+M) $\phi(a_1a_2\dots a_{n-k})(\alpha_{n-k+1}\dots\alpha_N)$ must be completely symmetric in both sets of indices a_i or α_i since the original indices $(A_1\dots A_n)$ were completely symmetrized. Thus, it transforms as the direct product representation $(\overbrace{\square\square\square\square\square\square}^{n-k}, \overbrace{\square\square\square\square\square}^{k})$ of SU(N) x SU(M). But for SU(N/M), since supersymmetrization of $(A_1A_2\dots A_n)$ implies symmetrization of the bosons ϕ_{a_i} and antisymmetrization of the fermions ψ_{α_i}, the bosonic indices $(a_1a_2\dots a_{n-k})$ are symmetric but the fermionic indices $(\alpha_{n-k+1}\dots\alpha_n)$ are antisymmetric. Thus, from (186) we can write the branching rule

$$\overbrace{\square\square\square\square\square\square\square\square}^{n} = \sum_{k=0} (\overbrace{\square\square\square\square\square}^{n-k}, \overbrace{\square\square\square}^{k})_{(n-k)/N - k/M} \qquad \text{for SU(N+M)} \tag{187}$$

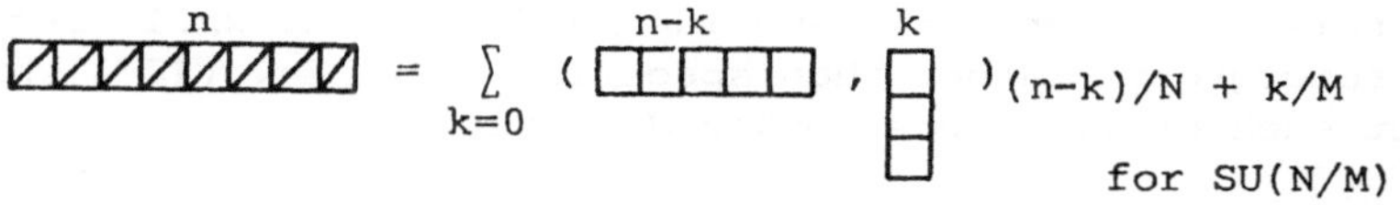
n
n-k
k
k=0
(n-k)/N + k/M
for SU(N/M)

The number of terms appearing in the first of these equations is (n+1). However, in the second equation some of the Young tableaux  k for SU(M) vanish if k > M. Thus, if n ⩽ M there will be n+1 terms, but if n > M there will be fewer terms.

Note that the pictures in eq. (187) are independent of the value of N and M. They are completely determined by the permutation symmetry of the original 1 row tableau. Therefore, the comparison of the branching rule for Lie groups and Lie supergroups need not be restricted to groups that have identical subgroups. For example, instead of comparing the branching rules of SU(6/4) to those of SU(10) we may just as well compare them to those of SU(75), since the pictures of SU(10) and SU(75) are identical except that Young tableaux for SU(10) with more than 10 rows vanish. Similarly, for a given shape of the supertableau the pictures of SU(6/4) are identical with those of any other SU(N/M) except for the illegal SU(N) or SU(M) tableaux that vanish. Thus, in branching rule calculations, we will always consider N and M to be as large as necessary so that none of the SU(N) or SU(M) tableaux vanish for both SU(N+M) and SU(N/M). Thus, the branching rule pictures that are obtained will be generic to the tableau and independent of the values of N and M. They will depend only on the number m_i and n_i of the dotted and undotted boxes in the original tableau as e.g. in Fig. 1. After obtaining the branching rule for a given tableau specified by m_i and n_i (not for given M and N), we can specialize to any desired values of N and M and eliminate, if necessary, any illegal SU(N) or SU(M) tableau. A legal SU(N) tableau which contains dotted boxes is specified in Fig. 2. Note that the number of dotted plus undotted boxes cannot exceed N

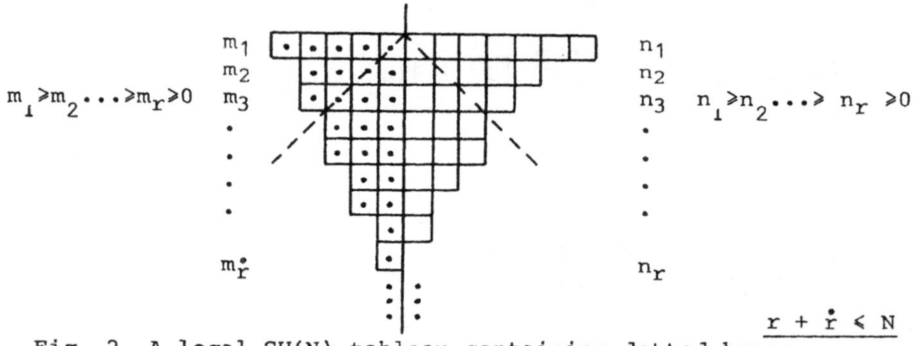

Fig. 2 A legal SU(N) tableau containing dotted boxes

In this way we obtain branching rules for the whole series of SU(N/M) supergroups rather than specific N, M. Eq. (187) which is the first such example, was obtained in ref. [14].

Next we consider the SU(N+M) tableau 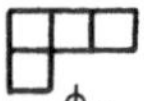 and SU(N/M) super-tableau 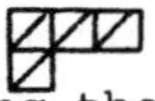 which correspond to a tensor $\phi_{(A_1A_2A_3);B_1}$. By specializing the indices $A_i = a_i \oplus \alpha_i$, $B_1 = b_1 \oplus \beta_1$, we obtain the various components. The <u>independent</u> components are specified by considering the indices a_i to be <u>lower</u> than the indices b_i within SU(N) and both a_i and b_i to be lower than the α_i or β_i within SU(N+M), when they are allowed to take values A=1,2,...,N+M. Then we obtain

$$\begin{aligned}\phi_{(A_1A_2A_3);B_1} &= \phi_{(a_1a_2a_3);b_1} \oplus \phi_{(a_1a_2a_3);\beta_1} \oplus \phi_{(a_1a_2)(\alpha_3);b_1} \\ &\oplus \phi_{(a_1a_2)(\alpha_3);\beta_1} \oplus \phi_{(a_1)(\alpha_2\alpha_3);b_1} \oplus \phi_{(a_1)(\alpha_2\alpha_3);\beta_1} \\ &\oplus \phi_{(\alpha_1\alpha_2\alpha_3);\beta_1} \quad .\end{aligned} \tag{188}$$

Note that we did not include $\phi_{(\alpha_1\alpha_2\alpha_3);b_1}$, and some others, even though they could appear as possible components of the tensor. This is because of the ordering rule $a_i<b_i<\alpha_i<\beta_i$ which allows us to select the independent components of the tensor <u>only once</u>. According to this rule, we cannot allow b_1 to appear in the second row when the first row contains only α's. The component with the symmetries of $\phi_{(\alpha_1\alpha_2\alpha_3);b_1}$ is already counted as seen below.

For SU(N+M) eq. (188) can be written in terms of SU(N) x SU(M) Young tableaux as

$\phi_{(a_1a_2a_3);b_1}$ = (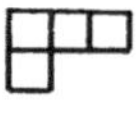 , 1) , $\phi_{(a_1a_2a_3);\beta_1}$ = ( , )

$\phi_{(a_1a_2)(\alpha_3);b_1}$ = (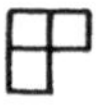 , ) , $\phi_{(a_1a_2)(\alpha_3);\beta_1}$ = ( , )

$\oplus$ (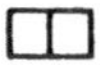 , ) ,

(189)

$\phi_{(a_1)(\alpha_2\alpha_3);b_1}$ = (,) , $\phi_{(a_1)(\alpha_2\alpha_3);\beta_1}$ = (, )

$\oplus$ (,)

$$\phi_{(\alpha_1\alpha_2\alpha_3);\beta_1} = (1, \boxbox)$$

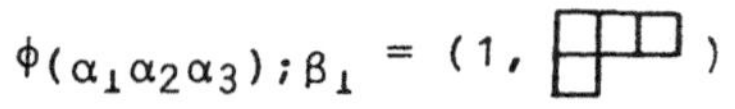

Note that $\phi_{(a_1a_2)(\alpha_3);\beta_1}$ has two irreducible SU(M) components since the α_3 index in the first row and the β_1 index in the second row are not forced to be in any symmetry relation relative to each other. Thus, we obtain the two irreducible pieces, because for SU(M),

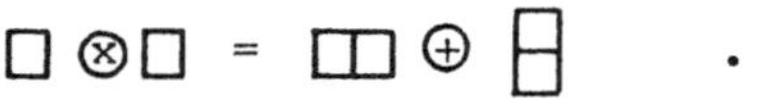

$$\square \otimes \square = \square\square \oplus \begin{array}{c}\square\\\square\end{array} \quad . \tag{190}$$

Similarly, $\phi_{(a_1)(\alpha_2\alpha_3);\beta_1}$ has two irreducible components since

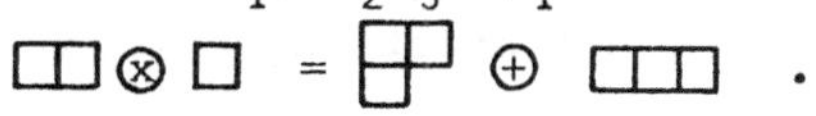

$$\square\square \otimes \square = \boxbox \oplus \square\square\square \quad . \tag{191}$$

The second piece in eq. (191) corresponds to the component $\phi_{(\alpha_1\alpha_2\alpha_3);\beta_1}$ mentioned above.

Thus, for SU(N+M) $\supset$ SU(N) x SU(M) x U(1) the branching rule for this tableau is

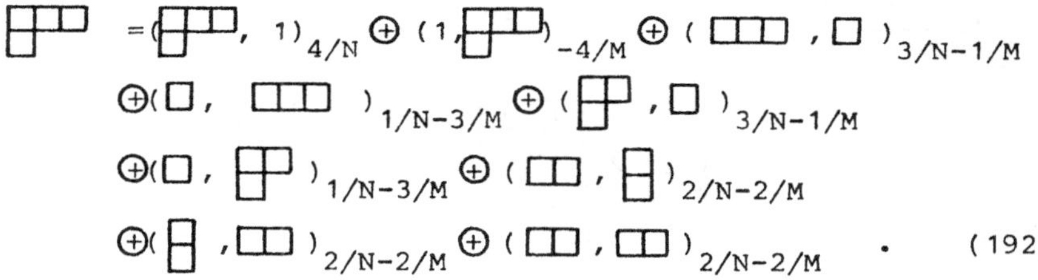

$$\boxbox = (\boxbox, 1)_{4/N} \oplus (1, \boxbox)_{-4/M} \oplus (\square\square\square, \square)_{3/N-1/M}$$
$$\oplus(\square, \square\square\square)_{1/N-3/M} \oplus (\boxbox, \square)_{3/N-1/M}$$
$$\oplus(\square, \boxbox)_{1/N-3/M} \oplus (\square\square, \begin{array}{c}\square\\\square\end{array})_{2/N-2/M}$$
$$\oplus(\begin{array}{c}\square\\\square\end{array}, \square\square)_{2/N-2/M} \oplus (\square\square, \square\square)_{2/N-2/M} \quad . \tag{192}$$

Note that in the final result the <u>pictures</u> are symmetric under the interchange of SU(M) and SU(N), as they should be, since the permutation symmetry of the original tableau does not distinguish between SU(N) and SU(M) indices. Identical results would be obtained by considering the a_i,b_i indices to be higher compared to the α_i indices. The SU(N) $\leftrightarrow$ SU(M) interchangeability of the branching rule reflects this fact. Thus, for every irreducible component (X,Y) there exists another irreducible component (Y,X) where X and Y represent the <u>pictures</u> of Young tableaux.

The same reasoning can be applied step by step to the SU(N/M) group. The only difference is that whenever the α_i's were symmetrized within SU(N+M) they should be antisymmetrized within SU(N/M) and viceversa. This is required by the supersymmetrization indicated by the supertableau. The a_i's have the same permutation properties in both SU(N+M) and SU(N/M) as in the previous example. Thus, for SU(N/M), eq. (189) will be modified by changing every SU(M) row into a column and vice-versa. This means that every irreducible component (X,Y) that appeared for SU(N+M) will have a counterpart (X,$\tilde{Y}$) for SU(N/M) where $\tilde{Y}$ is an SU(M) Young tableau reflected along the diagonal relative to Y. Thus, the analog of eq. (192)

for $SU(N/M) \supset SU(N) \times SU(M) \times U(1)$ becomes

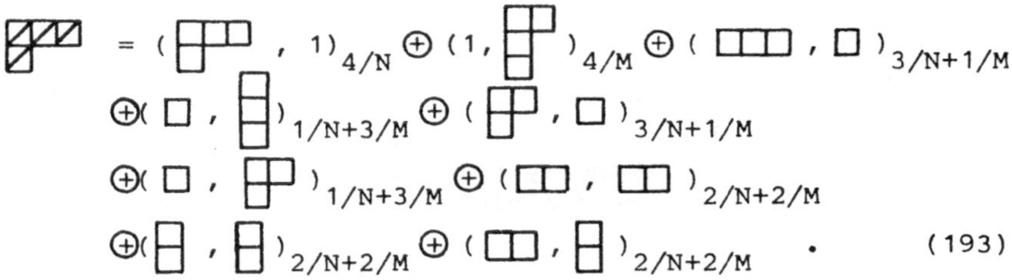

$$\young = (\;,\;1)_{4/N} \oplus (1,\;)_{4/M} \oplus (\;,\;)_{3/N+1/M}$$
$$\oplus (\;,\;)_{1/N+3/M} \oplus (\;,\;)_{3/N+1/M}$$
$$\oplus (\;,\;)_{1/N+3/M} \oplus (\;,\;)_{2/N+2/M}$$
$$\oplus (\;,\;)_{2/N+2/M} \oplus (\;,\;)_{2/N+2/M} \quad . \qquad (193)$$

Note that the 1/M pieces in the U(1)'s have switched signs. Furthermore the pictures are now symmetric under the $SU(N) \longleftrightarrow SU(M)$ interchange only after being reflected along the diagonal. That is, for every irreducible component $(X, \widetilde{Y})$ there exists also $(Y, \widetilde{X})$.

We use the same methods as above for tableaux containing n_1 boxes in the first row and n_2 boxes in the second row, where $n_1 \geqslant n_2 \geqslant 0$. The result for $SU(N+M)$ is

$$\begin{matrix} n_1 \\ n_2 \end{matrix} \;\boxed{}\; = \sum_{k_2=0}^{n_2} \sum_{j=0}^{k_2} \sum_{k_1=k_2-j}^{n_1-n_2+k_2-j}$$

$$\left[\begin{matrix} n_1-k_1 \\ n_2-k_2 \end{matrix}\; \boxed{}\; , \; \begin{matrix} k_1+j \\ k_2-j \end{matrix} \;\boxed{} \right] ,$$

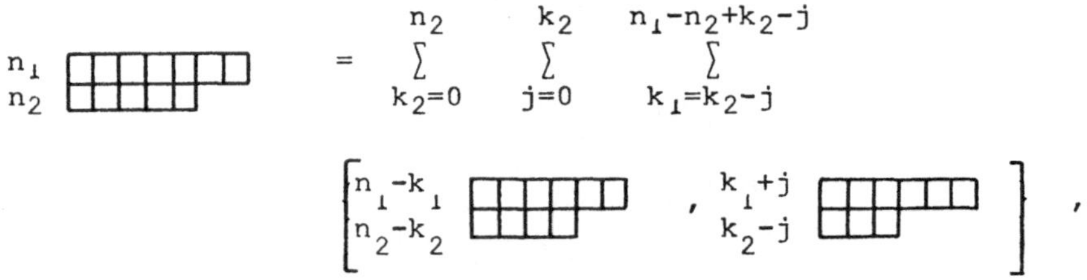

$$(194)$$

while for $SU(N/M)$ we only need to reflect the $SU(M)$ tableau and obtain

$$\begin{matrix} n_1 \\ n_2 \end{matrix} \;\boxed{}\; = \sum_{k_2=0}^{n_2} \sum_{j=0}^{k_2} \sum_{k_1=k_2-j}^{n_1-n_2+k_2-j}$$

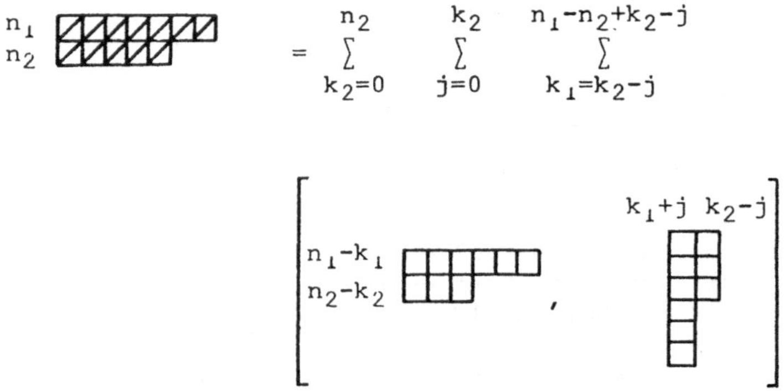

$$(195)$$

The U(1) charges were not indicated for lack of space but they are computed by counting the number of boxes for SU(N) and SU(M) tableaux and given as $(n_1+n_2-k_1-k_2)/N \mp (k_1+k_2)/M$, with the upper sign for SU(N+M) and the lower one for SU(N/M). The number of independent irreducible terms appearing in the sum of eqs. (194-195) is easily computed to be

$$1/2\ (n_1-n_2+1)(n_2+1)(n_2+2) \quad . \tag{196}$$

Some of these terms may vanish if N or M are too small and the tableau becomes illegal. Equations (194-195) reduce to (18) for the special case $n_2=0$.

After obtaining the result for 1 and 2 rows, it is inmmediate to arrive at the branching rule for 1 and 2 columns, simply by reflecting each tableau along the diagonal. These and a few other cases not containing dotted boxes are summarized in Table 1.

Returning to tableaux containing dotted boxes, we begin with the fundamental contravariant representation, $\phi^A=\phi^a \oplus \psi^a$, which may be written as

$$\boxdot = (\boxdot\ ,\ 1)_{-1/N} \oplus (1,\ \boxdot\ )_{1/M} \quad \text{for SU(N+M)} \quad ,$$

$$\tag{197}$$

$$\boxslash = (\boxdot\ ,\ 1)_{-1/N} \oplus (1,\ \boxdot\ )_{-1/M} \quad \text{for SU(N/M)} \quad .$$

The next simplest case is the adjoint representation $\phi_A{}^B$ which is a traceless matrix for SU(N+M) and supertraceless for SU(N/M). Specializing the indices $A=a \oplus \alpha$, $B=b \oplus \beta$, we obtain the components

$$\phi_A{}^B = \phi_a{}^b \oplus \phi_a{}^\beta \oplus \phi_\alpha{}^b \oplus \phi_\alpha{}^\beta \quad . \tag{198}$$

The pieces $\phi_a{}^b$ and $\phi_\alpha{}^\beta$ are not irreducible with respect to SU(N) and SU(M) respectively, since we have not yet insured that they are traceless. Thus

$$\phi_a{}^b = \tilde{\phi}_a{}^b + \frac{1}{N}\,\delta_a{}^b\ \phi = (\boxdot\square\ ,\ 1) \oplus (1,1) \quad ,$$

$$\tag{199}$$

$$\phi_\alpha{}^\beta = \tilde{\phi}_\alpha{}^\beta \mp \frac{1}{M}\,\delta_\alpha{}^\beta\ \phi = (1,\ \boxdot\square\ ) \oplus (1,1) \quad ,$$

where the singlet part $\phi=(1,1)$ is identical in both pieces so that the tracelessness (supertracelessness) condition on $\phi_A{}^B$ is satisfied. The result for SU(N+M) is

$$\boxdot\square \ = (\boxdot\square\ ,\ 1)_0 \oplus (1,\ \boxdot\square\ )_0 \oplus (1,1)_0 \oplus (\square\ ,\ \boxdot\ )_{1/N+1/M}$$

$$\oplus (\boxdot\ ,\ \square\ )_{-1/N-1/M} \quad , \tag{200}$$

TABLE 1. Branching Rules for SU(N+M) and SU(N/M)

SU(N+M) / SU(N/M) IRREP	SU(N) ⊗ SU(M) IRREP	U(1) CHARGE	LIMITS ON NUMBERS OF BOXES	TOTAL NUMBER OF TERMS IF ALL TABLEAUX ARE LEGAL
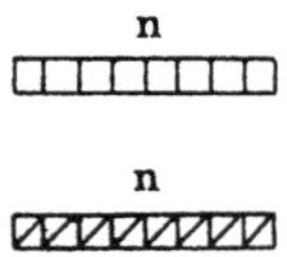	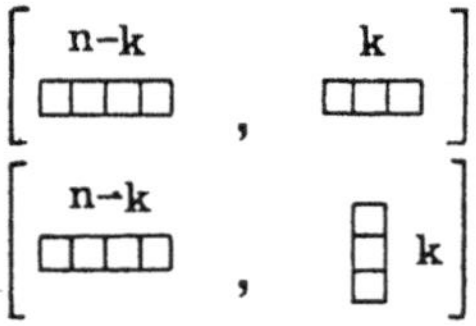	$\frac{n-k}{N}+\frac{k}{M}$	$0 \leq k \leq n$	$n+1$
	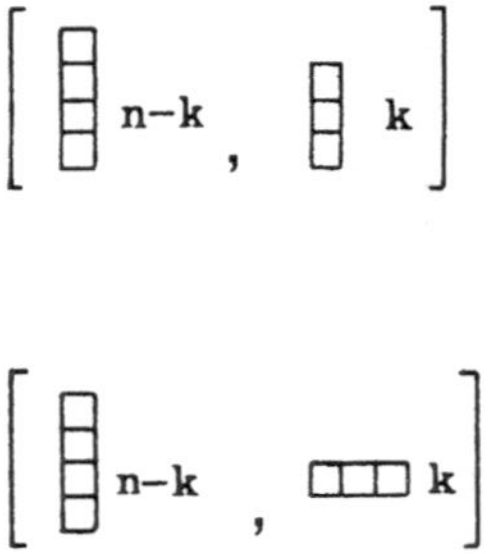	$\frac{n-k}{N}+\frac{k}{M}$	$0 \leq k \leq n$	$n+1$
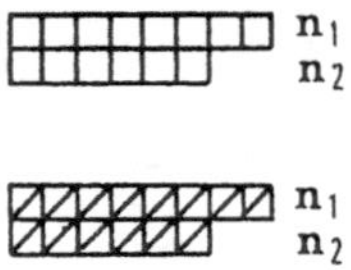	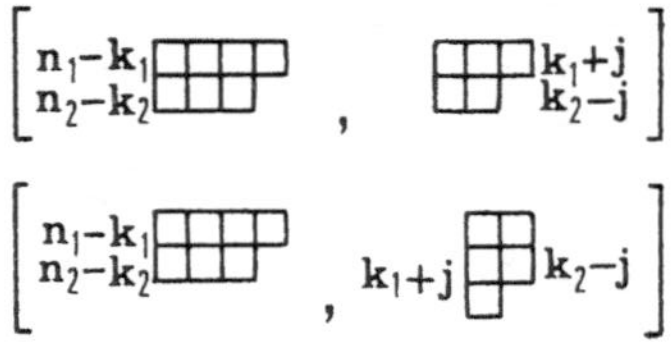	$\frac{n_1+n_2-k_1-k_2}{N}+\frac{k_1+k_2}{M}$	$0 \leq k_2 \leq n_2$ $0 \leq j \leq k_2$ $k_2-j \leq k_1 \leq n_1-n_2+k_2-j$	$1/2\,(n_1-n_2+1)(n_2+1)\times$ $\times(n_2+2)$

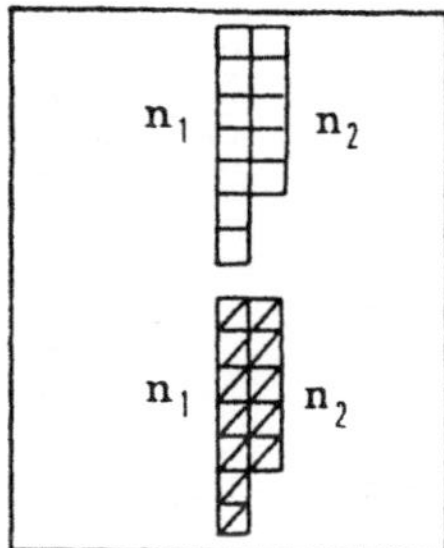	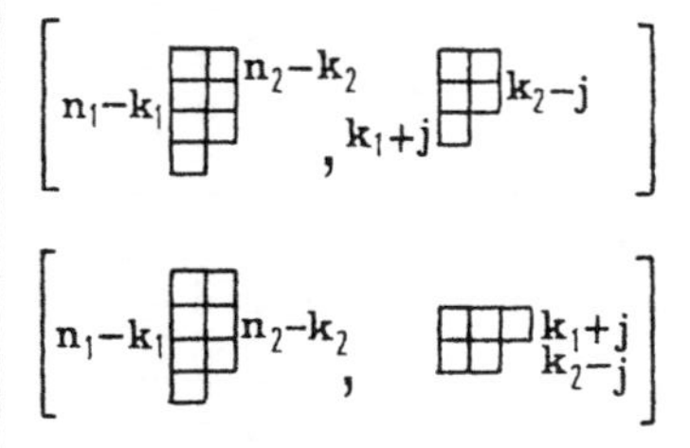	$\frac{n_1+n_2-k_1-k_2}{N} - \frac{k_1+k_2}{M}$ (with $+$ below the $-$)	$0 \leq k_2 \leq n_2$ $0 \leq j \leq k_2$ $k_2-j \leq k_1 \leq n_1-n_2+k_2-j$	$1/2\,(n_1-n_2+1)(n_2+1)(n_2+2)$
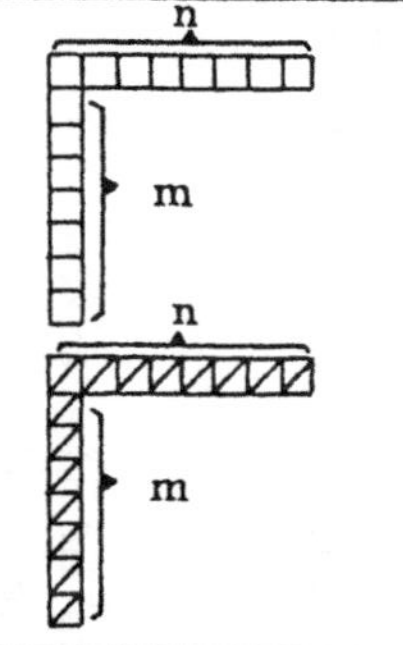	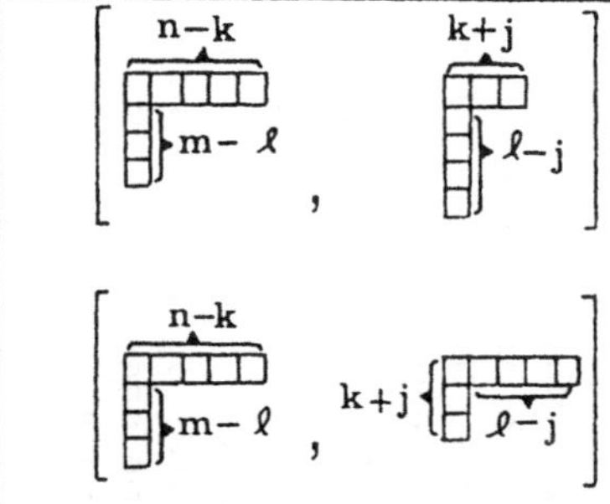	$\frac{n+m-k-\ell}{N} - \frac{k+\ell}{M}$ (with $+$ below the $-$)	$j=0,1$ $j \leq \ell \leq m$ $1-j-\delta_{\ell_0} \leq k \leq n-1+(1-j)\,\delta_{\ell_m}$ δ_{ℓ_i} = Kronecker delta	$2nm+n-m+1$
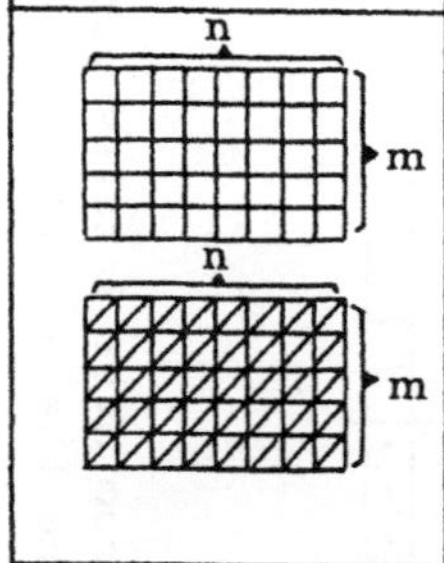	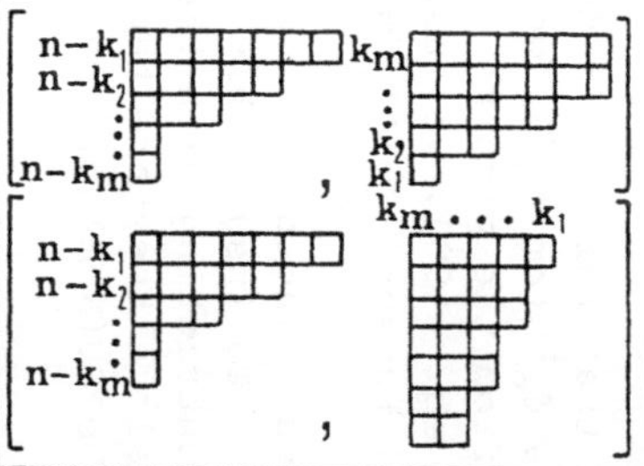	$\frac{nm}{N} - \left[\frac{1}{N} + \frac{1}{M}\right]\left[\sum_{i=1}^{m} k_i\right]$	$0 \leq k_1 \leq k_2 \; \ldots \leq k_m \leq n$	$\frac{(n+m)!}{n!\,m!}$

while for SU(N/M) we have

$$\boxed{/}\boxed{/} = (\boxed{\cdot}\boxed{},\ 1)_0 \oplus (1,\ \boxed{\cdot}\boxed{})_0 \oplus (1,1)_0 \oplus (\boxed{},\ \boxed{\cdot})_{1/N-1/M}$$

$$\oplus(\boxed{\cdot},\ \boxed{})_{-1/N+1/M} \quad . \tag{201}$$

Next we consider the tensor $\phi_{A_1A_2\cdots A_n}^{B_1B_2\cdots B_m}$ with both lower and upper indices symmetrized (or supersymmetrized) and satisfying the trace (or superstrace) condition. Specializing the indices, we have

$$\phi_{A_1A_2\cdots A_n}^{B_1B_2\cdots B_m} = \sum_{k=0}^{m} \sum_{\ell=0}^{m} \phi_{(a_1a_2\cdots a_{n-k})(\alpha_{n-k+1}\cdots\alpha_n)}^{(b_1b_2\cdots b_{n-\ell})(\beta_{m-\ell+1}\cdots\beta_m)} \tag{202}$$

The U(1) quantum numbers can be calculated by assigning 1/N for each a_i, -1/N for each b_i, ($\mp$ 1/M) for each α_i and ($\pm$ 1/M) for each β_i, where the upper sign is for SU(N+M) and the lower sign for SU(N/M). The various terms in the sum are, in general, reducible with respect to SU(N) $\otimes$ SU(M). For example, for n=2 and m=2,

$$\phi_{(a_1a_2)}^{(b_1b_2)} = (\boxed{\cdot}\boxed{\cdot}\boxed{}\boxed{},\ 1) \oplus (\boxed{\cdot}\boxed{},\ 1) \oplus (1,\ 1) \quad ,$$

$$\phi_{(a_1\alpha_2)}^{(b_1b_2)} = (\boxed{\cdot}\boxed{\cdot}\boxed{},\ \boxed{}) \oplus (\boxed{\cdot},\ \boxed{}),\ \text{etc.} \tag{203}$$

Note that for each reduction which is achieved by using a Kronecker delta $\delta_a{}^b$ (or $\delta_\alpha{}^\beta$), we get to eliminate one dotted and one undotted box from the picture of an SU(N) (or an SU(M)) tableau. Just as the ϕ in eq. (199), we must be aware that the tracelessness of the original tensor $\phi_{A_1\cdots A_n}^{B_1\cdots B_m}$ imposes that some of the pieces in the various traces are identical and should not be counted more than once. To insure this property we count only the traces calculated by contracting with the SU(M) $\delta_\alpha{}^\beta$, and ignore those obtained with $\delta_a{}^b$ since they are the same ones. With these conditions, we arrive at the SU(N+M) branching rule

$$\overbrace{\boxed{\cdot}\boxed{\cdot}\boxed{\cdot}\boxed{\cdot}\boxed{\cdot}}^{m}\overbrace{\boxed{}\boxed{}\boxed{}\boxed{}\boxed{}}^{n} = \sum_{k=0}^{n} \sum_{\ell=0}^{m} \sum_{i=0}^{\min(k,\ell)} \left[\overbrace{\boxed{\cdot}\boxed{\cdot}\boxed{\cdot}\boxed{\cdot}}^{m-\ell}\overbrace{\boxed{}\boxed{}\boxed{}}^{n-k},\ \overbrace{\boxed{\cdot}\boxed{\cdot}\boxed{\cdot}}^{\ell-i}\overbrace{\boxed{}\boxed{}\boxed{}}^{k-i} \right] . \tag{204}$$

The sum over i takes care of the traces and produces te pieces similar to the $\phi \sim (1,1)$ of eq. (199). The U(1) charge, which depends only on the number of boxes, is given as $(n-k-m+\ell)/N-(k-\ell)/M$.

For $SU(N/M)$ the reasoning is identical, however, the α_i's and β_i's should now be antisymmetrized as opposed to being symmetrized in the previous case. Therefore, the $SU(M)$ tableaux should be changed relative to the previous case by substituting columns instead of rows. Otherwise, every step can be repeated to obtain

$$
\underbrace{\boxed{\diagdown\diagdown\diagdown\diagdown}}_{m}\;\underbrace{\boxed{\diagup\diagup\diagup\diagup}}_{n} = \sum_{k=0}^{n}\sum_{\ell=0}^{m}\sum_{i=0}^{\min(k,\ell)}\left[\underset{m-\ell\quad n-k}{\boxed{\bullet\;\bullet\;\bullet\;\bullet}}\quad,\quad \ell-i\;\boxed{\begin{matrix}\bullet\\\bullet\\\bullet\end{matrix}}\;k-i\right].
$$

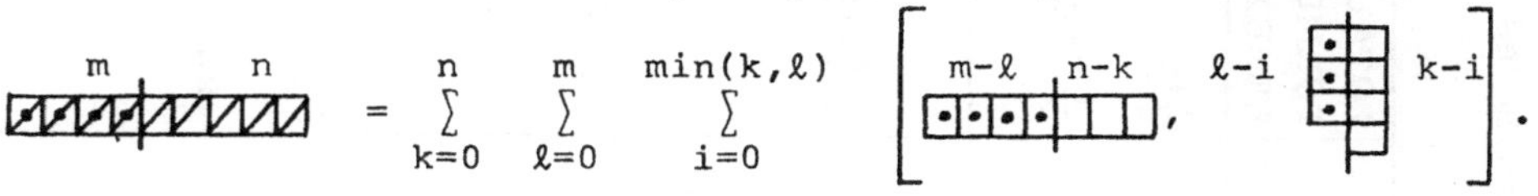

(205)

We see that in going from $SU(N+M)$ to $SU(N/M)$, the $SU(M)$ tableaux get reflected independently for the dotted and undotted boxes along their respective diagonals. The diagonals are shown in fig. 2.

Our method should be quite clear to the reader by now. Without giving any more details, we list our results for a few tableaux containing dotted boxes in table 2. We emphasize that these correspond to arbitrarily large representations of arbitrarily large groups. Together with table I we expect that these concrete results should be quite sufficient for a variety of physical applications that we can now foresee. More complicated cases can be worked out, if necessary, with the same methods.

16. **Branching Rules for $SU(N_1+N_2/M_1+M_2) \supset SU(N_1/M_1) \otimes SU(N_2/M_2) \otimes$** **$U(1)$**

This branching is again obtained from that of $SU(N+M) \supset SU(N) \times SU(M) \times U(1)$ by a reinterpretation of the boxes in the tableaux. Let us first identify the fundamental representation $\phi_A = \phi_a \oplus \phi_\alpha$ as follows: ϕ_a contains N_1 bosons and M_1 fermions and belongs to the fundamental representation $\phi_a \sim \boxed{\diagup}$ of $SU(N_1/M_1)$; similarly, ϕ_α contains N_2 bosons and M_2 fermions and belongs to the fundamental representation $\phi_\alpha \sim \boxed{\diagup}$ of $SU(N_2/M_2)$.

The $U(1)$ generator, which is a $N_1+M_1+N_2+M_2$ dimensional diagonal matrix in the fundamental representation, is identified as

$$
\left[\begin{array}{c|c} \dfrac{1}{N_1 - M_1} & 0 \\ \hline 0 & \dfrac{-1}{N - M} \end{array}\right]\begin{array}{l}\left.\vphantom{\dfrac{1}{N}}\right\}\; N_1 + M_1 \\[2ex] \left.\vphantom{\dfrac{1}{N}}\right\}\; N_2 + M_2\end{array}\quad,
$$

TABLE 2. Branching Rules for SU(N+M) Irreps Containing Dotted Boxes

SU (N+M) / SU (N/M) IRREP	SU(N) ⊗ SU(M) IRREP	U(1) CHARGE	LIMITS ON NUMBERS OF BOXES	TOTAL NUMBER OF TERMS IF ALL TABLEAUX ARE LEGAL
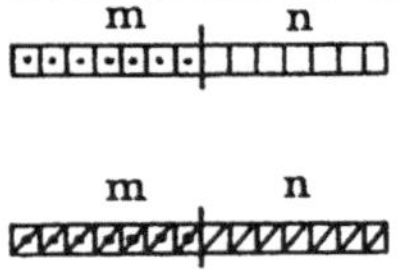	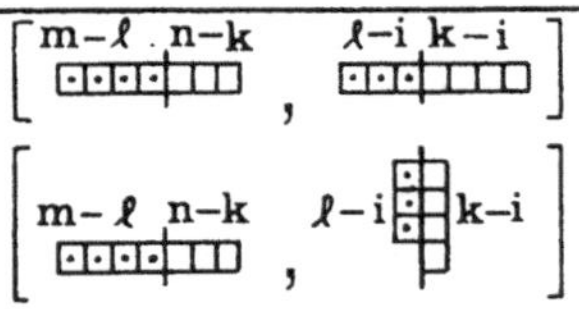	$\frac{n-k-m+\ell}{N} \mp \frac{k-\ell}{M}$	$0 \leq k \leq n$ $0 \leq \ell \leq m$ $0 \leq i \leq \min(k,\ell)$	$\frac{1}{6}(m+1)(m+2)(3n-m+3)$ if $n \geq m$ $\frac{1}{6}(n+1)(n+2)(3m-n+3)$ if $n \leq m$
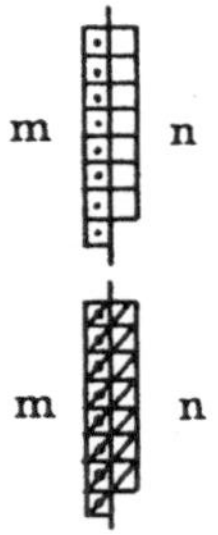	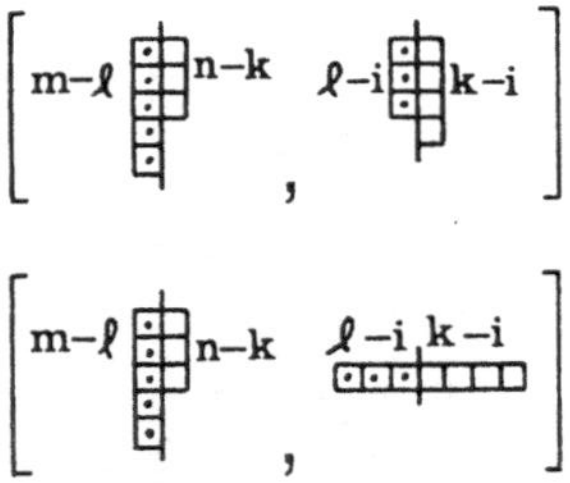	$\frac{n-k-m+\ell}{N} \mp \frac{k-\ell}{M}$	$0 \leq k \leq n$ $0 \leq \ell \leq m$ $0 \leq i \leq \min(k,\ell)$	$\frac{1}{6}(m+1)(m+2)(3n-m+3)$ if $n \geq m$ $\frac{1}{6}(n+1)(n+2)(3m-n+3)$ if $n \leq m$
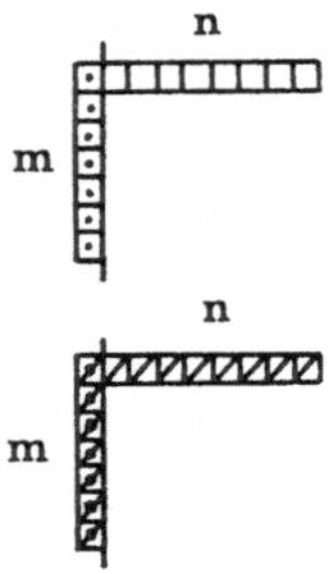	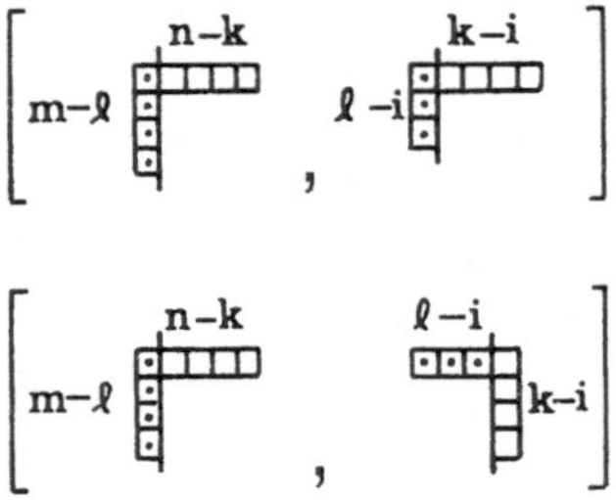	$\frac{n-k-m+\ell}{N} \mp \frac{k-\ell}{M}$	$i=0,1$ $i \leq k \leq n$ $i \leq \ell \leq m$	$2mn+m+n+1$

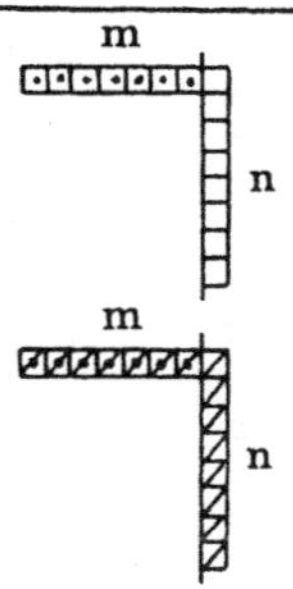	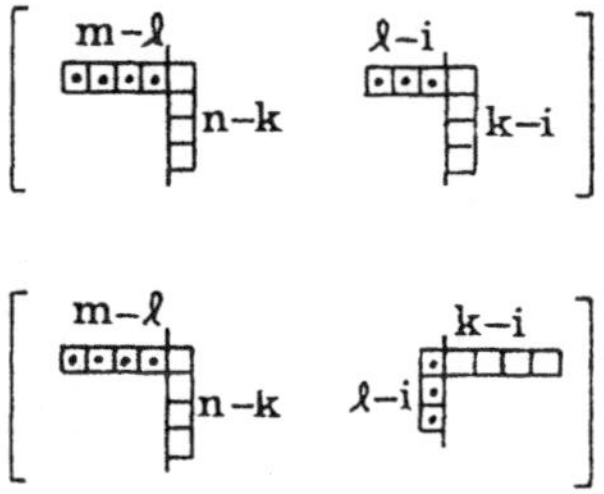	$\frac{n-k-m+\ell}{N} \mp \frac{k-\ell}{M}$	$i = 0,1$ $i \leqslant k \leqslant n$ $i \leqslant \ell \leqslant m$	$2mn+m+n+1$
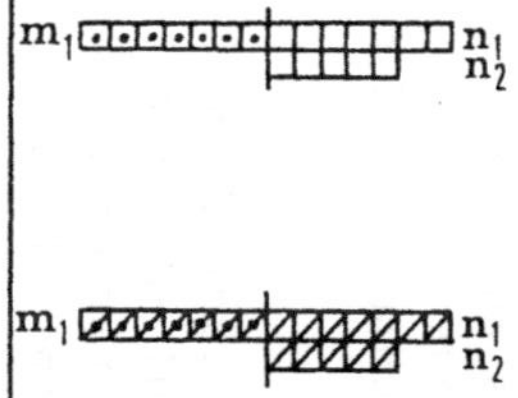	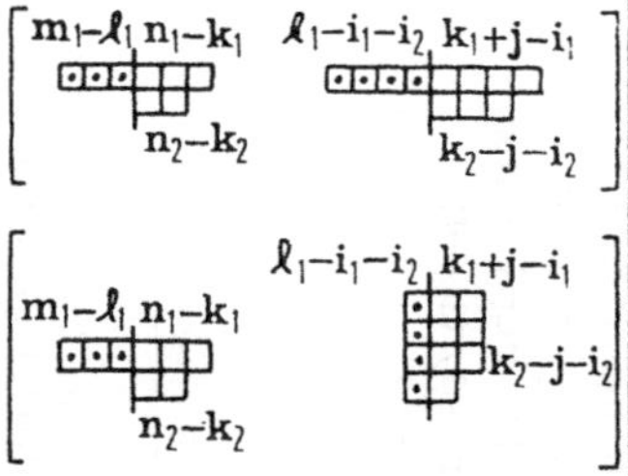	$\frac{n_1+n_2-m_1}{N} -$ $- \left[\frac{1}{N} \pm \frac{1}{M}\right]\left[k_1+k_2-\ell_1\right]$	$0 \leqslant \ell_1 \leqslant m_1$ $0 \leqslant k_2 \leqslant n_2$ $0 \leqslant j \leqslant k_2$ $k_2-j \leqslant k_1 \leqslant n_1-n_2+k_2-j$ $0 \leqslant i_2 \leqslant \min(k_2-j, \ell_1)$ $0 \leqslant i_1 \leqslant \min(k_1-k_2+2j, \ell_1-i_2)$	
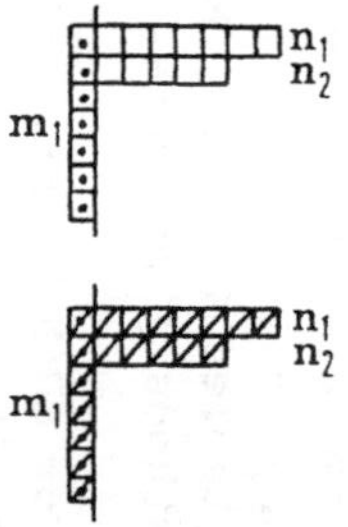	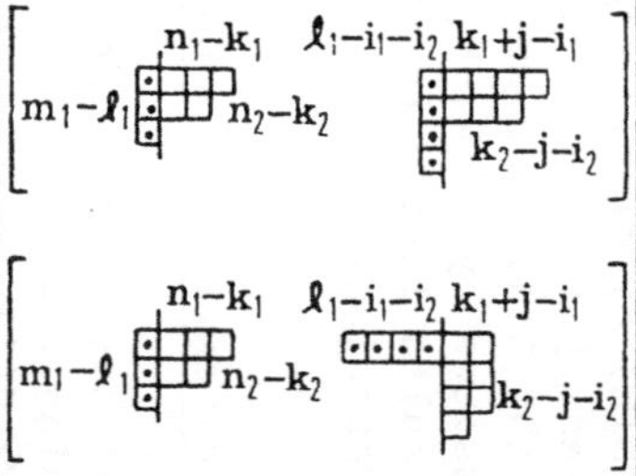	$\frac{n_1+n_2-m_1}{N} -$ $- \left[\frac{1}{N} \mp \frac{1}{M}\right]\left[k_1+k_2-\ell_1\right]$	$i_1 = 0,1\ ;\ i_2 = 0,1$ $i_1+i_2 \leqslant \ell_1 \leqslant m_1$ $i_2 \leqslant k_2 \leqslant n_2$ $0 \leqslant j \leqslant k_2-i_2$ $i_1-i_2+k_2-j \leqslant k_1 \leqslant n_1-n_2+k_2-j$	$\frac{1}{2}(n_2+1) \times$ $\times \left[(3m_1n_2+2)(n_1-n_2) + 2m_1(2n_1+1)+2\right]$ $m_1 \geq 1$

Thus, for each index a or α we obtain the U(1) charges $1/(N_1-M_1)$ or $-1/(N_2-M_2)$ respectively. Therefore, $\phi_A = \phi_a \oplus \phi_\alpha$ may be written in terms of tableaux as

$$\boxslash = (\boxslash\,,\,1)_{1/(N^1-M^1)} \oplus (1,\boxslash\,)_{-1/(N^2-M^2)} \quad . \tag{207}$$

Note the formal similarity to eq. (185) for SU(N+M) except that every box is replaced by a slashed box, and the values of the U(1) charges are computed by different assignments to a and α, as explained above.

The completely supersymmetric tensor $\phi(A_1 A_2 \ldots A_n)$ can be decomposed by specializing each index $A_i = a_i \oplus \alpha_i$, just as in eq. (186). However, the meaning of $\phi(a_1, a_2 \ldots a_n)$ etc., now differs from the SU(N+M) case in that the indices $(a_1 a_2 \ldots)$ or $(\alpha_1 \alpha_2 \ldots)$ are supersymmetrized rather than simply symmetrized. Therefore, eq. (187) now gets replaced by

$$\overset{n}{\boxslash\boxslash\boxslash\boxslash\boxslash\boxslash\boxslash} = \sum_{k=0}^{n} \Big(\overset{n-k}{\boxslash\boxslash\boxslash\boxslash}\,,\; \overset{k}{\boxslash\boxslash\boxslash} \Big) \quad , \tag{208}$$

with the U(1) charges given by $\left[(n-k)/(N_1-M_1) - k/(N_2-M_2)\right]$.

The analysis is the same for any other supertableau and the result for the new branching rules is obtained from the known cases of SU(N+M) by simply replacing every SU(N) or SU(M) box $\square$ by slashed boxes $\boxslash$ belonging to $SU(N_1/M_1)$ or $SU(N_2/M_2)$ respectively. Similarly the U(1) is computed by replacing every 1/N or -1/M in the old expressions by $1/(N_1-M_1)$ or $-1/(N_2-M_2)$ respectively. Therefore, all the results listed in tables 1 and 2 are directly generalized to the new branching.

17. Comments on Branching Rules

We have established a one-to-one correspondence between the branching rules $SU(N+M) \supset SU(N) \otimes SU(M) \otimes U(1)$ and $SU(N/M) \supset SU(N) \otimes SU(M) \otimes U(1)$ as well as $SU(N_1+N_2/M_1+M_2) \supset SU(N_1/M_1) \otimes SU(N_2/M_2) \otimes U(1)$. Some concrete examples which we expect to be sufficient for most physical applications have been explicitly worked out and listed in tables 1 and 2. More complicated cases can be analyzed with the methods given here.

For SU(N+M) branching rules, there are useful lists available in the literature[22]. Our SU(N+M) results are in complete agreement with these known cases. One virtue of our approach is that we are not limited by large dimensions of representations or groups. Thus, in our tables 1 and 2 one finds large dimensions not covered in the extensive lists mentioned.

In making comparisons with these lists one must be aware that some Young tableaux for SU(N) or SU(M) become illegal (see fig. 2) and vanish if either N or M are too small. One may use the available SU(N+M) lists to derive additional practical SU(N/M) results not covered in this paper explicitly, provided N and M are large enough to insure no tableau vanishes. Useful branching rules can be obtained from ref. [22] or similar lists by noting the following general observations which follow from our analysis above.

Consider an arbitrary Young tableau T for SU(N+M) and the SU(N) $\otimes$ SU(M) $\otimes$ U(1) branching rule

$$T = \sum \oplus (X,Y) \quad , \tag{209}$$

where X and Y denote Young tableaux for SU(N) and SU(M) respectively. In general, T,X,Y contain both dotted and undotted boxes. In tables such as ref. [13], Dynkin indices are used to label a representation. They must be converted to Young tableau notation in order to apply our method below.

For every SU(N+M) Young tableau T, as in fig. 1, we can define an SU(N/M) supertableau $\not T$, with identical numbers n_i, m_i for its rows, except that every box $\square$ or $\boxdot$ is replaced by a slashed box $\boxslash$ or $\boxslash$. We may then consider the branching rule for SU(N/M) $\rightarrow$ SU(N) $\otimes$ SU(M) $\otimes$ U(1) as

$$\not T = \sum \oplus (X,\tilde{Y}) \quad , \tag{210}$$

where $\tilde{Y}$ is the reflection of Y along its diagonals. The diagonals are shown in fig. 2. Similarly, the branching rule for $SU(N_1+N_2/M_1+M_2) \supset SU(N_1/M_1) \otimes SU(N_2/M_2) \otimes U(1)$ can be written as

$$\not T = \sum \oplus (\not X,\not Y) \quad , \tag{211}$$

where $\not X$ and $\not Y$ are supertableaux analogous to X and Y. Equations (210) and (211) follow from (209) if one uses the tensor language $\phi_{A_1A_2\ldots}^{B_1B_2\ldots}$ and specializes each index $A_i = a_i \oplus \alpha_i$, etc. It is necessary to consider the meaning of supersymmetrization in relation to ordinary symmetrization, and then extracting irreducible SU(N) and SU(M) components. These statements can be understood by following the examples in tables (1,2).

Equation (209) contains the terms (1,T) and (T,1) where T is the SU(N) or SU(M) representation with the most boxes that could appear in the branching rule. The pictures that represent the decomposition $T = \sum \oplus (X,Y)$ are independent of N or M. Therefore, we will assume that N and M are large enough so that the Young tableau T or the reflection from its diagonals $\tilde{T}$ does not vanish for SU(N) or SU(M). This insures that every (X,Y) or $(X,\tilde{Y})$ that could appear in the sum does not vanish for SU(N+M) or SU(N/M). After obtaining the branching rules for such large N,M we can apply the result to smaller N,M, as it may be necessary in some practical application. Then, we only need to eliminate the illegal SU(N) or SU(M) tableaux according to fig. 2.

The following statements hold for any decomposition noted in eqs. (209), (210), (211), and can be used as a check in any calculation.

(1) The total number of undotted minus dotted boxes is identical in every term of the sum and equal to the same quantitiy for T or $\not{T}$.

(2) For a term (X,Y) that appears in eq. (209) there should be another term (Y,X) provided it does not vanish according to fig. 2. For eq. (210) this implies that for every term $(X,\tilde{Y})$ there should be a $(Y,\tilde{X})$, while in eq. (211) for every $(\not{X},\not{Y})$ there should exist a $(\not{Y},\not{X})$.

(3) The maximum number or rows and columns that can appear in any (X,Y) or $(X,\tilde{Y})$ or $(\not{X},\not{Y})$ as well as the general shape of these Young tableaux are predetermined by the number of rows and columns and general shape of T or $\not{T}$. See the examples in tables 1, 2.

(4) The dimension of $\not{T}$ or T on the left-hand side of these equations should match with the sum of the dimensions on the right-hand side. For this purpose one can use the practical dimension formulas for the numbers of bosons and fermions developed in Ref. [14] and in previous sections. On the right-hand side of eq. (210) a fermion is obtained when the SU(M) representation $\tilde{Y}$ contains an odd total number of dotted plus undotted boxes (independent of X). (This is a fermion in a class I representation. For class II we demand an odd number of boxes in X rather than in Y. While in mixed class I - class II representations, the roles of bosons and fermions may be interchanged. The representation of the group element is independent from the class).

18. Branching Rules $SU(N_1N_2+M_1M_2/N_1M_2+M_1N_2) \rightarrow SU(N_1/M_1) \otimes SU(N_2/M_2)$

This branching is analogous to SU(NM) ---> SU(N) x SU(M), and as in the previous section we will use such analogies. Consider the fundamental bases of $SU(N_1/M_1)$, $\Phi_{A_1} = (\phi_{a_1},\psi_{\alpha_1}),(a_1 = 1,\ldots,N_1;$ $\alpha_1=1,2,\ldots M_1)$

and of $SU(N_2/M_2)$, $\Phi'_{A_2} = (\phi'_{a_2}, \psi'_{\alpha_2})$,

with $(a_2=1,2,\ldots N_2;\ \alpha_2=1,2,\ldots M_2)$.

Take their product and count the number of bosons and fermions

$$\begin{aligned} &\text{bosons:} \quad \phi_{a_1}\phi'_{a_2} \oplus \psi_{\alpha_1}\psi'_{\alpha_2} \dashrightarrow N_1N_2 + M_1M_2 \ , \\ &\text{fermions:} \quad \phi_{a_1}\psi'_{\alpha_2} \oplus \psi_{\alpha_1}\phi_{a_2} \dashrightarrow N_1M_2 + M_1N_2 \ . \end{aligned} \qquad (212)$$

We assign these bosonic and fermionic states to the fundamental basis of $SU(N_1N_2+M_1M_2/N_1M_2+N_2M_1)$. Then we can write the branching rule

$$\boxslash = (\boxslash\,,\boxslash) \quad , \qquad (213)$$

which simply expresses what we have just described. Here the first box describes $SU(N_1/M_1)$ and the second one describes $SU(N_2/M_2)$.

To obtain the branching rules for higher dimensional representations we take direct products and supersymmetrize. For example

$$\boxslash\boxslash = \left[\,(\boxslash\,,\boxslash)\otimes(\boxslash\,,\boxslash)\,\right]_S \quad , \qquad (214)$$

where S stands for symmetric product under superpermutations. A superpermutation is one in which wavefunctions are permuted but indices kept fixed in the same order, as in Eq. (102). Another way of describing it is a permutation of the indices in a tensor, provided $\pm$ signs are inserted depending on whether a pair of bosonic or fermionic indices are permuted, as in Eq. (104). In any case, superpermutations are built-in in our notation of supertableaux. The pictures that describe a symmetric product under superpermutations, such as $\boxslash\boxslash$, look the same as the ordinary symmetric Young tableau except for the slashes which summarize the supersymmetrization rule. This notation, of course, extends to the direct product. The result will look the same as in the case of $SU(NM) \dashrightarrow SU(N)\times SU(M)$ except for the slashes through the boxes which automatically take into account the superpermutation rule. Thus, we obtain

$$\boxslash\boxslash = (\boxslash\boxslash\,,\boxslash\boxslash) + \left(\begin{smallmatrix}\boxslash\\ \boxslash\end{smallmatrix}\,,\begin{smallmatrix}\boxslash\\ \boxslash\end{smallmatrix}\right) \quad , \qquad (215)$$

where each term in the sum is symmetric according to super permutations. Similarly, for the antisymmetric product the result is

$$\begin{smallmatrix}\boxslash\\ \boxslash\end{smallmatrix} = \left(\begin{smallmatrix}\boxslash\\ \boxslash\end{smallmatrix}\,,\boxslash\boxslash\right) + \left(\boxslash\boxslash\,,\begin{smallmatrix}\boxslash\\ \boxslash\end{smallmatrix}\right) \quad , \qquad (216)$$

where each term is antisymmetric under superpermutations.

It is now straightforward to obtain the branching for higher representations by simply copying the pictures that would apply to SU(NM) ---→ SU(N) x SU(M), except for the slashes. Thus, we write a few cases

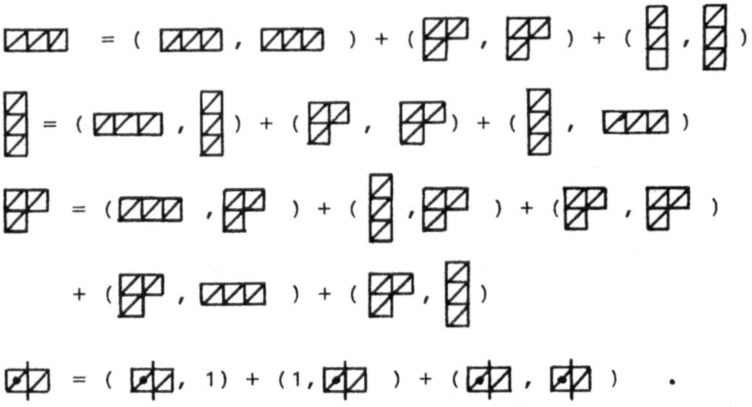

(217) (218) (219) (220)

In general, for a completely symmetric row we have

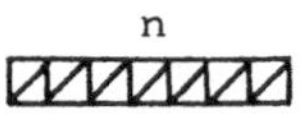

$$= \sum_{Y} (\not{Y}n, \not{Y}n) \quad , \tag{221}$$

where on the right hand side both entries have identical-looking supertableaux $\not{Y}_n$ with n boxes in each, and the sum extends over all possible n-box tableaux. Similarly for a column we have

$$= \sum_{Y} (\not{Y}n \ , \tilde{\not{Y}}n) \quad , \tag{222}$$

where the second entry $\tilde{\not{Y}}n$ is reflected along the diagonal relative to the first entry $\not{Y}n$, and the sum extends over all supertableaux containing n-boxes.

Of course, as in the case of SU(N/M) ---→ SU(N) x SU(M) x U(1), available tables[22] for ordinary groups can be used to arrive at the branching of more complicated supertableaux, provided that the result is first expressed in terms of Young tableaux, and one insures that M, N were large enough so that none of the pictures that would have entered have been eliminated. After this step, by simply replacing each box by a superbox we arrive at the correct branching for the supergroup. The pictures that describe such branchings, provided they are complete, are inherent to the tableau and do not depend on the size of the group (i.e., N_1, M_1, N_2, M_2) so that the branchings for arbitrarily large supergroups and representations are readily obtained in this way.

19. Kac-Dynkin Diagrams and Supertableaux

We begin by recalling the relation between ordinary Young tableaux and Dynkin diagrams. Any of the states described by a Young tableau is assigned a weight or a collection of "charges". The weight is given by the eigenvalues of the commuting Cartan generators (commuting charge operators) on that state. In the fundamental representation of SU(N) the N-1 commuting generators are chosen as

$$H_1 = \begin{bmatrix} 1 & & & & & & & \\ & -1 & & & & & & \\ & & 0 & & & & & \\ & & & 0 & & & & \\ & & & & 0 & & & \\ & & & & & \cdot & & \\ & & & & & & \cdot & \\ & & & & & & & \cdot \end{bmatrix}, \quad H_2 = \begin{bmatrix} 0 & & & & & & & \\ & 1 & & & & & & \\ & & -1 & & & & & \\ & & & 0 & & & & \\ & & & & 0 & & & \\ & & & & & \cdot & & \\ & & & & & & \cdot & \\ & & & & & & & \cdot \end{bmatrix}, \tag{223}$$

$$H_3 = \begin{bmatrix} 0 & & & & & & & \\ & 0 & & & & & & \\ & & 1 & & & & & \\ & & & -1 & & & & \\ & & & & 0 & & & \\ & & & & & \cdot & & \\ & & & & & & \cdot & \\ & & & & & & & \cdot \end{bmatrix}, \quad \ldots \; H_{N-1} = \begin{bmatrix} 0 & & & & & & & & \\ & 0 & & & & & & & \\ & & \cdot & & & & & & \\ & & & \cdot & & & & & \\ & & & & \cdot & & & & \\ & & & & & \cdot & & & \\ & & & & & & 0 & & \\ & & & & & & & 1 & \\ & & & & & & & & -1 \end{bmatrix}.$$

Then, their eigenvalues on the states $\square \sim \phi_i$, $i=1,2,\ldots N$, are computed via matrix multiplication. For example, if we take the basis

$$\boxed{i} \sim |i\rangle \sim \begin{bmatrix} 0 \\ 0 \\ \cdot \\ \cdot \\ \cdot \\ 1 \\ 0 \\ \cdot \\ \cdot \\ \cdot \\ 0 \end{bmatrix} \longleftarrow \text{i'th entry} \quad , \tag{224}$$

then

$$H_\ell |i\rangle \sim \sum_j (H_\ell)_i{}^j |j\rangle \quad , \tag{225}$$

inmediately yields the eigenvalues. Some examples are given:

$$H_1\,|1\rangle = 1|1\rangle\,,\; H_2|1\rangle = 0|1\rangle\,,\; H_3|1\rangle = 0|1\rangle,\; \text{etc.}\,,$$

$$H_1\,|2\rangle = -1|2\rangle\,,\; H_2|2\rangle = 1|2\rangle\,,\; H_3|2\rangle = 0|2\rangle,\; \text{etc.}\,, \tag{226}$$

$$H_1\,|3\rangle = 0|3\rangle\,,\; H_2|3\rangle = -1|3\rangle\,,\; H_3|3\rangle = 1|3\rangle,\; \text{etc.}\;.$$

So, for any state, a collection of eigenvalues of H_ℓ, such as $(h_1,h_2\cdots h_{N-1})$, can be given. Thus, we write the weights corresponding to the states of the fundamental representation

$$\begin{aligned} |1\rangle &\dashrightarrow (1,0,0,0,\ldots,0)\,,\\ |2\rangle &\dashrightarrow (-1,1,0,0,\ldots,0)\,,\\ |3\rangle &\dashrightarrow (0,1,-1,0,\ldots,0)\,,\\ &\;\vdots\\ |N\rangle &\dashrightarrow (0,0,0,\ldots,0,-1)\;. \end{aligned} \tag{227}$$

The weights are given an order such that the eigenvalues of H_1 are specified first, next the eigenvalues of H_2, etc., all the way to H_{N-1}.

The highest weight is the one that has the largest <u>positive</u> eigenvalues first relative to H_1, then H_2, then H_3, etc. Thus, in the fundamental representation the highest weight is

$$\square \dashrightarrow (1,0,0,\ldots,0) \tag{228}$$

and the highest state is then 1 .

It is easy to compute the highest weight of a higher representation in the tensor notation or in terms of Young tableaux. For example for the symmetric tensor $\phi_{ij} \sim \boxed{i\,|\,j}$ the highest state corresponds to i=1, j=1 $\to \phi_{11} \sim \boxed{1\,|\,1}$ and the highest weight is then simply obtained by applying H_ℓ to each index i=1, j=1, giving

$$\square\!\square \dashrightarrow (2,0,0,\ldots,0)\;. \tag{229}$$

This is computed by applying the matrices $(H_\ell)_i{}^j$ on ϕ_{ij} in the form of an infinitesimal transformation (adding the charges)

$$(H_\ell\,\phi)_{ij} \equiv (H_\ell)_i{}^{i'}\,\phi_{i'j} + (H_\ell)_j{}^{j'}\,\phi_{ij'}\,, \tag{230}$$

which inmediately yields (2,0,0,...0) on ϕ_{11} ~ 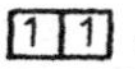. Similarly for the n'th order symmetric tensor $\phi_{i_1 i_2 \ldots i_n}$ ~ 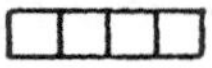 the highest state is specified by $i_1 = i_2 = \ldots = i_n = 1$, and the highest weight is

(n,0,0,...,0) ~  . (231)

The eigenvalues in the highest weight coincide with the Dynkin indices $(a_1, a_2 \ldots, a_{N-1})$ for SU(N). Thus, we can establish a one-to-one correspondence between Young tableaux and Dynkin diagrams via the highest weight as in Eqs. (228, 229, 231)

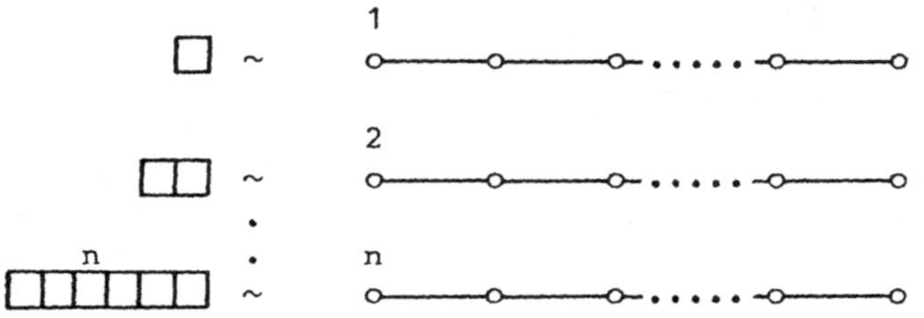

(232)

Generalizing to more complicated Young tableaux it can be seen that for undotted boxes the highest state is found by assigning i=1 to every box on the first row, i=2 to every box on the second row, i=3 to every box on the third row etc.

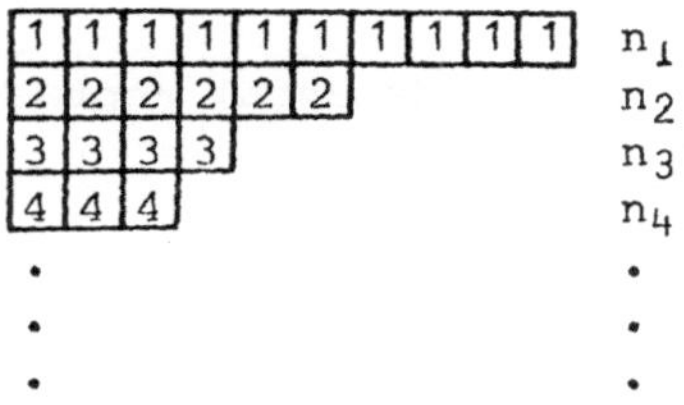

(233)

The eigenvalues of the "charges" $H_1, H_2, H_3, \ldots$, etc. on this state are

$$(n_1 - n_2, n_2 - n_3, n_3 - n_4, \ldots) \quad . \tag{234}$$

This is so since H_1 assigns +1 to $|1\rangle$ and −1 to $|2\rangle$, zero to all other states; similarly H_2 picks nonzero charges from $|2\rangle$ and $|3\rangle$, etc. This provides the translation form Young tableaux to Dynkin diagrams. Thus for the tableau of Eq. (233) the Dynkin diagram is

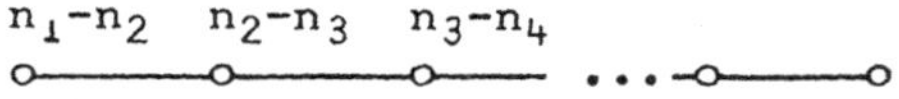

(235)

If the tableau has only dotted boxes, then the "charges" (H_ℓ) are applied with the opposite sign on the states of the tensor with upper indices (antiparticles). Then a little thought will show that the highest state is obtained by assigning the values of indices i=N to every box in the first row, i=N-1 to every box in the second row, etc.

$$\begin{array}{l|c|c|c|c|c|c|c|}
\cline{2-8}
m_1 & \cdot N & \cdot N & \cdot N & \cdot N & \cdot N & \cdot N & \cdot N \\
\cline{2-8}
m_2 & \multicolumn{2}{c|}{} & \cdot N-1 & \cdot N-1 & \cdot N-1 & \cdot N-1 & \cdot N-1 \\
\cline{4-8}
m_3 & \multicolumn{4}{c|}{} & \cdot N-2 & \cdot N-2 & \cdot N-2 \\
\cline{6-8}
\cdot & \multicolumn{5}{c|}{} & \cdot N-3 & \cdot N-3 \\
\cline{7-8}
\cdot & \multicolumn{5}{c}{} & \multicolumn{1}{c}{\cdot} & \multicolumn{1}{c}{\cdot} \\
\cdot & \multicolumn{5}{c}{} & \multicolumn{1}{c}{\cdot} & \multicolumn{1}{c}{\cdot}
\end{array} \qquad (236)$$

Then the highest weight, computed by applying H_ℓ on this state (with the opposite sign!), is

$$(0,0,\ldots,m_3-m_4,\ m_2-m_3,\ m_1-m_2) \quad . \qquad (237)$$

Finally, a general tableau of the type of Fig. (2) has a highest state specified by the assignments of Eqs. (233) and (236) on the undotted and dotted boxes. So its highest weight is

$$(n_1-n_2,\ n_2-n_3,\ \ldots\ ,\ m_2-m_3,\ m_1-m_2) \qquad (238)$$

and the Dynkin diagram corresponding to Fig. (2) is then

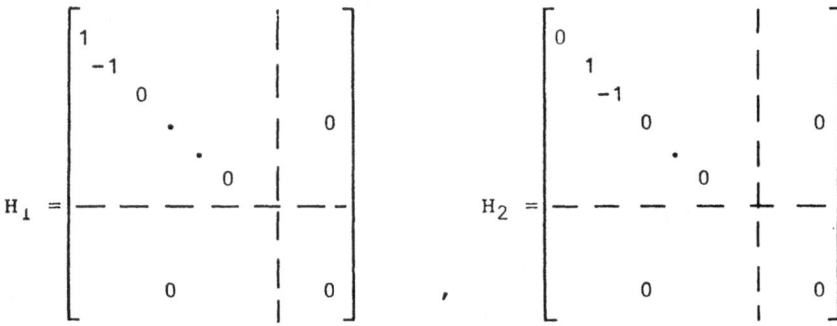

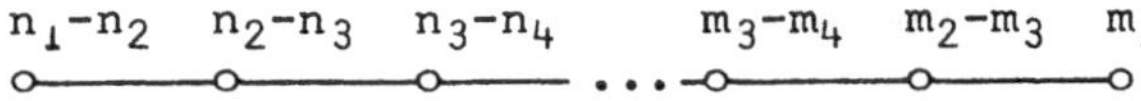

(239)

Now, we return to supergroups. In the fundamental representations the commuting Cartan matrices are chosen as

$$H_1 = \left[\begin{array}{cccccc|c}
1 & & & & & & \\
 & -1 & & & & & \\
 & & 0 & & & & \\
 & & & \cdot & & & 0 \\
 & & & & \cdot & & \\
 & & & & & 0 & \\
\hline
 & & & & & & \\
 & & 0 & & & & 0
\end{array}\right] , \qquad
H_2 = \left[\begin{array}{cccccc|c}
0 & & & & & & \\
 & 1 & & & & & \\
 & & -1 & & & & \\
 & & & 0 & & & 0 \\
 & & & & \cdot & & \\
 & & & & & 0 & \\
\hline
 & & & & & & \\
 & & & 0 & & & 0
\end{array}\right]$$

$$
,\ldots, \quad H_{n-1} = \left[\begin{array}{cccccc:c}
0 & & & & & & \\
& \cdot & & & & & \\
& & \cdot & & & & \\
& & & 0 & & & 0 \\
& & & & 1 & & \\
& & & & & -1 & \\
\hdashline
& & & 0 & & & 0
\end{array}\right] ,
$$

$$
L_0 = \left[\begin{array}{ccccc:ccccc}
0 & & & & & & & & & \\
& \cdot & & & & & & & & \\
& & \cdot & & & & & 0 & & \\
& & & \cdot & & & & & & \\
& & & & 0 & & & & & \\
& & & & & 1 & & & & \\
\hdashline
& & & & & & 1 & & & \\
& & & & & & & 0 & & \\
& & 0 & & & & & & \cdot & \\
& & & & & & & & & \cdot \\
& & & & & & & & & \cdot \\
& & & & & & & & & 0
\end{array}\right] ,
\tag{240}
$$

$$
k_1 = \left[\begin{array}{c:cccccc}
0 & & & 0 & & & \\
\hdashline
& 1 & & & & & \\
& & -1 & & & & \\
0 & & & 0 & & & \\
& & & & \cdot & & \\
& & & & & \cdot & \\
& & & & & & 0
\end{array}\right] ,
$$

$$k_2 = \left[\begin{array}{c|cccccc} 0 & & & 0 & & & \\ \hline & 0 & & & & & \\ & & 1 & & & & \\ 0 & & & -1 & & & \\ & & & & 0 & & \\ & & & & & \cdot & \\ & & & & & & \cdot \end{array}\right] \quad , \ldots ,$$

$$k_{M-1} = \left[\begin{array}{c|cccccc} 0 & & & 0 & & & \\ \hline & 0 & & & & & \\ & & \cdot & & & & \\ 0 & & & \cdot & & & \\ & & & & 0 & & \\ & & & & & 1 & \\ & & & & & & -1 \end{array}\right] \quad .$$

Note that they must be supertraceless, and this fixes L_0 with its two positive entries in contrast to all the others.

Using the same approach as above, we compute the eigenvalues of these operators on the states of any representation. If we identify the highest state and highest weight by the same conventions as above, then the eigenvalues that specify the highest weight correspond to the indices in the Kac-Dynkin diagram

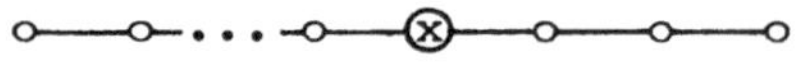 (241)

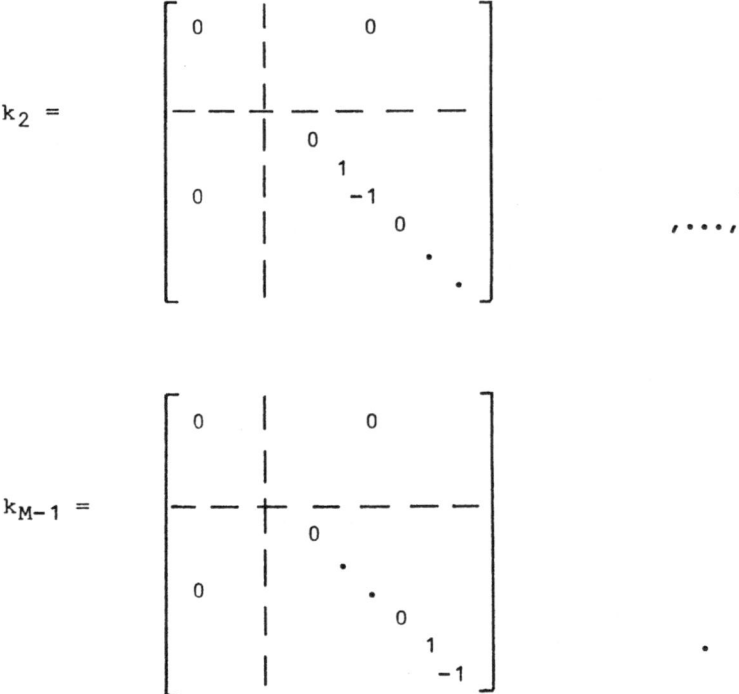

The circle with the cross corresponds to the eigenvalue of L_0.

To find the highest weight we analyze the SU(N) x SU(M) x U(1) content of any SU(N/M) representation as is done in sections (15-17), Tables 1, 2 and Eq. (210),

$$\not{R} = \sum (X, \tilde{Y})_q \quad , \tag{242}$$

where X, Y represent the pictures of SU(N) and SU(M) Young tableaux respectively, and q specifies the U(1) eigenvalues of all the states described by X, Y. Generally, we find that the highest state is contained in the $(X, \tilde{Y})$ component that carries the smallest value of q (in a convention $N > M$). q is given as

$$q = \frac{\alpha}{N} + \frac{\beta}{M} \quad , \tag{243}$$

where α is the number of undotted minus dotted boxes in X and β is the corresponding quantity in $\tilde{Y}$. Thus, to obtain the smallest value of q we must choose the $(X, \tilde{Y})$ component that has the most SU(N) undotted boxes in X and the most SU(M) dotted boxes in $\tilde{Y}$. This is easy to find pictorially directly from the original supertableaux as follows

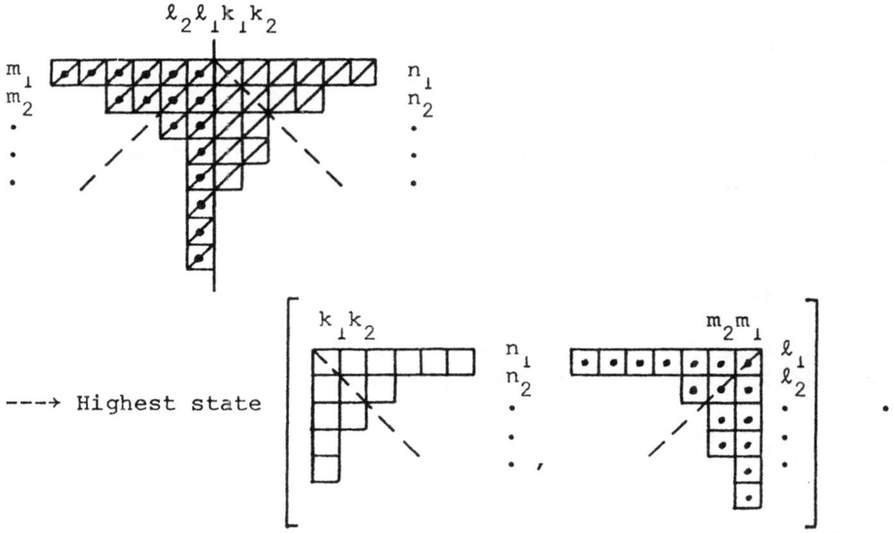

(244)

We simply have assigned all undotted boxes to SU(N) and all dotted boxes to SU(M) (recall that the SU(M) tableau is reflected along the diagonal relative to the original tableau). If N and M are sufficiently large so that this highest component $(X,\tilde{Y})$ is nonvanishing (i.e. they are legal SU(N) x SU(M) tableaux), then the highest state is contained here. The weight is now computed by applying the Cartan generators $H_1, H_2, \ldots H_{N-1}$; of Eq. (240). It is clear that the highest weight will then be L_0; $k_1, K_2, \ldots k_{M-1}$ of Eq. (240). It is clear that the highest weight will then be

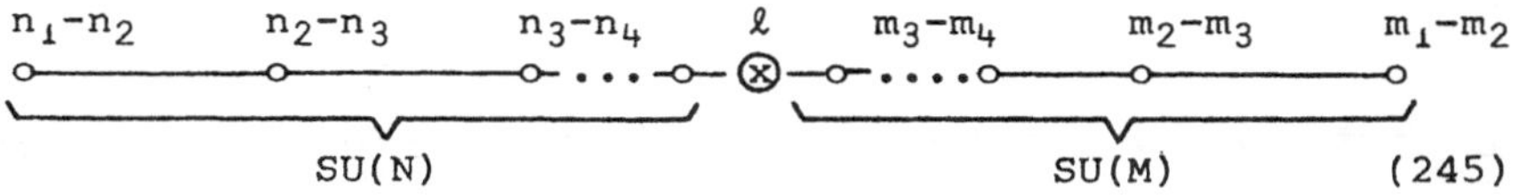

(245)

The eigenvalue ℓ clearly must be related to q (Eq. 243). We can

find the relation between q and ℓ through the fundamental representation, where the operator Q corresponding to the eigenvalue q is identified as the supertraceless matrix

$$Q = \left[\begin{array}{cccc|cccc} 1/N & & & & & & & \\ & 1/N & & & & & & \\ & & \ddots & & & & & \\ & & & 1/N & & & & \\ \hline & & & & 1/M & & & \\ & & & & & 1/M & & \\ & & & & & & \ddots & \\ & & & & & & & 1/M \end{array}\right] \quad . \tag{246}$$

Then we see that Q can be written in terms of the commuting Cartan generators as

$$Q = \sum_{i=1}^{N-1} \frac{i}{N} H_i - \sum_{j=1}^{M-1} \frac{j}{M} K_{M-j} + L_0 \quad . \tag{247}$$

Since we already know the eigenvalues of H_i, K_j, L_0 as specified in Eqs. (245), (243), we immediately deduce the value of ℓ from

$$\frac{\alpha}{N} + \frac{\beta}{M} = \sum_{i=1}^{N-1} \frac{i}{N}(n_i - n_{i+1}) - \sum_{j=1}^{M-1} \frac{j}{M}(m_j - m_{j+1}) + \ell \quad , \tag{248}$$

where α and β were given in Eq. (243). If the tableaux of Eq. (244) are non-vanishing, then α = number of boxes in SU(N) tableaux and β = number of boxes in SU(M) tableaux. Note that α, β must be computed before eliminating any completed columns which may have exactly N boxes for SU(N) and M boxes for SU(M).

Let us look at some examples, where we assume that N, M are very large so that the highest state is easily identified as in Eq. (244)

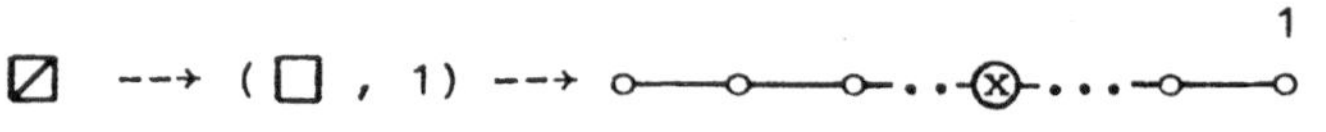

(249)

(250)

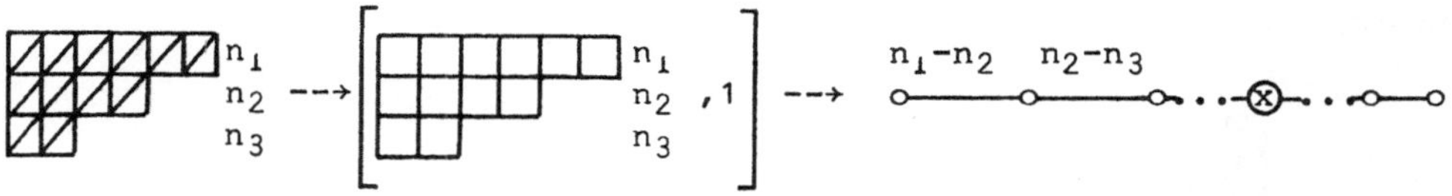

(251)

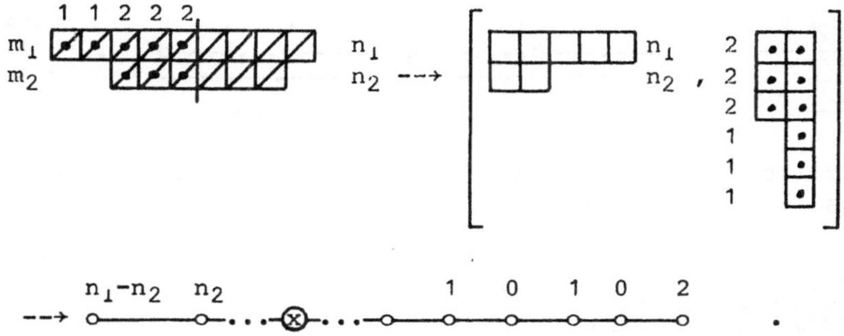

(252)

When there are a large number of undotted rows or a large number of dotted columns then the pictures of Eq. (244) become illegal. We must then return to the branching rules of the previous section, and Tables 1, 2, in order to identify the first nonvanishing component with the lowest value of the U(1) charge q. This is best explained with an example. Consider the supertableau

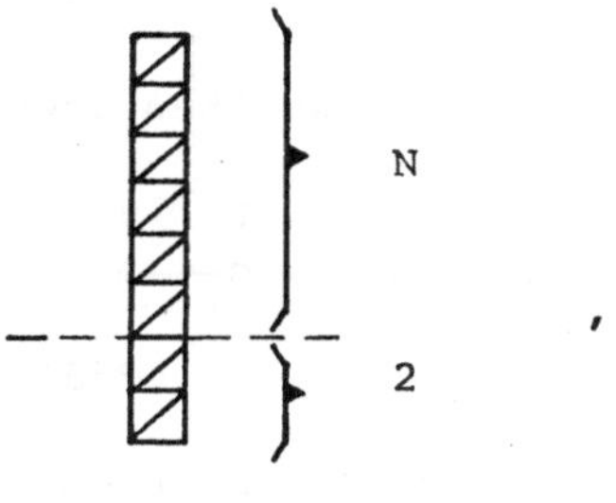

(253)

whose undotted boxes in a column exceed N. The simple prescription of Eq. (244) now fails. However, all we need to do, after the naive step of Eq. (244), is simply transfer the extra two boxes to SU(M) in order to find the first nonvanishing set of Young tableaux with the highest value of q. This is just

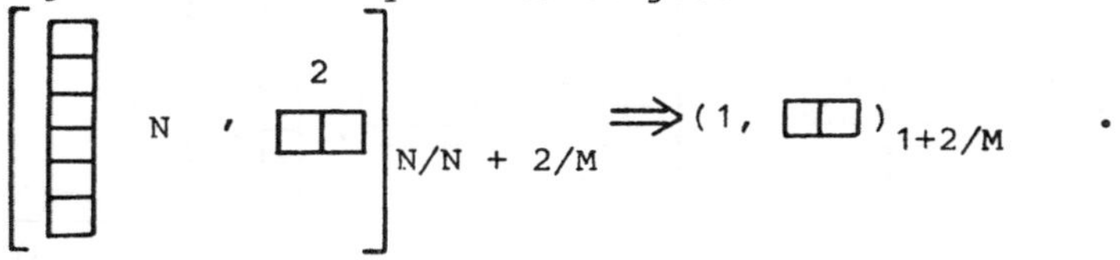

(254)

Note that the SU(M) tableaux are reflected relative to the original ones. Now the SU(N) representation is a singlet and the SU(M) representation is $\square\square$. Thus, the highest weight is identified and the

Kac-Dynkin diagram is

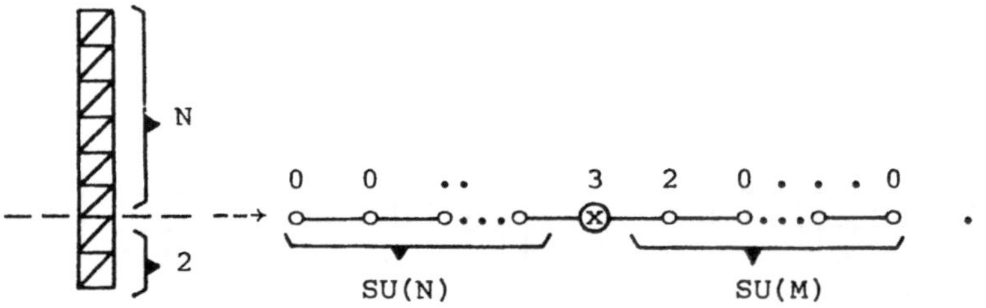

$$(255)$$

A similar situation would occur with a row of dotted super boxes whose length exceeds M

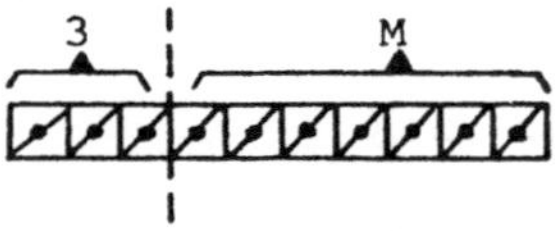

$$(256)$$

With arguments similar to the above, we identify the highest state as the first nonvanishing one with the smallest U(1) charge q. This is given by first considering the naive step of Eq. (244) and then transferring the extra 3 boxes to SU(N):

$$\left[\,\square\square\square\ \ ,\ \begin{array}{c}\boxdot\\\boxdot\\\boxdot\\\boxdot\\\boxdot\end{array}M\,\right]_{-3/N\,-\,M/M}\ \Rightarrow\ (\,\square\square\square\ ,\ 1)_{-3/N-1}$$

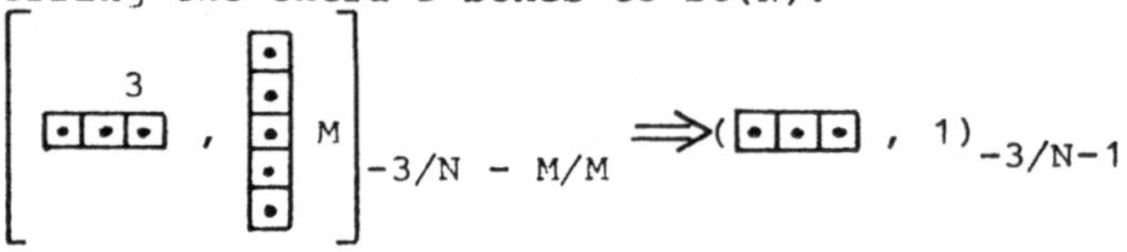

$$(257)$$

The highest weight and the Kac-Dynkin diagram is now written down

M+3

$$\boxed{\diagdown\diagup\diagup\diagup\diagup\diagup\diagup}\ \dashrightarrow\ \underset{0}{\circ}\ \underset{0}{\circ}\ \underset{0}{\circ}\ ..\ \underset{3}{\circ}\ \underset{-4}{\otimes}\ \underset{0}{\circ}\ ...\ \underset{0}{\circ}\!-\!\!\underset{0}{\circ}$$

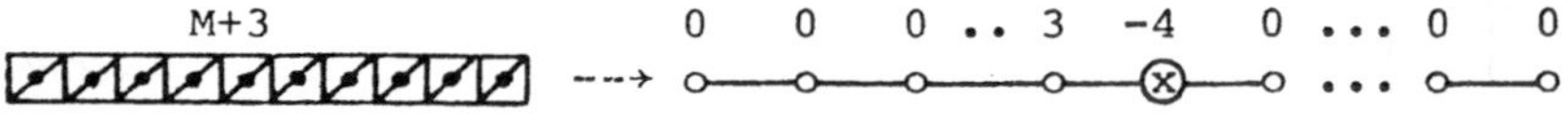

$$(258)$$

I list a few more examples without explanation, to be figured out by the reader.

Exercise 1

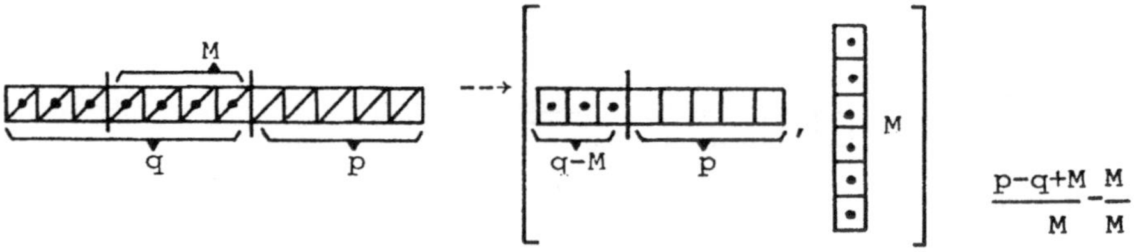

$$(259)$$

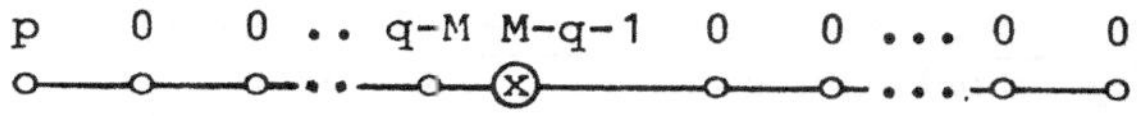
p 0 0 .. q-M M-q-1 0 0 ... 0 0

(260)

Exercise 2

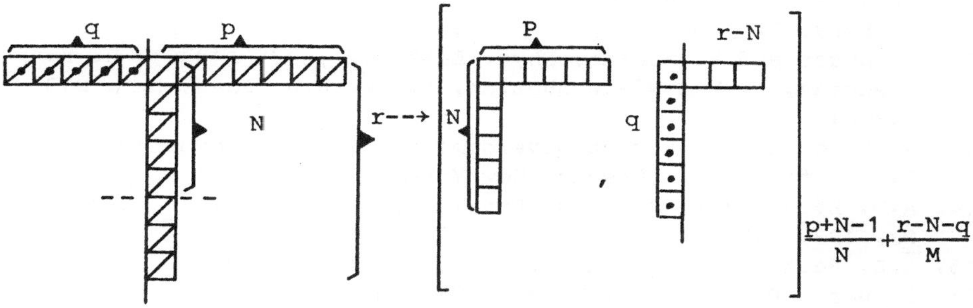
q
p
N
r
P
N
r-N
q
p+N-1 + r-N-q
N M

(261)

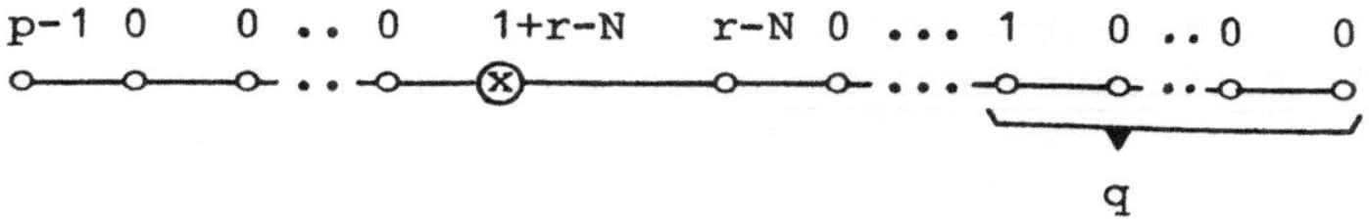
p-1 0 0 .. 0 1+r-N r-N 0 ... 1 0 .. 0 0
q

Further discussion can be found in a future publication.

References

1. A. Neveu and J.H. Schwarz, Nucl. Phys. B31, 86 (1971)
P. Ramond, Phys. Rev. D3, 2415 (1971).
2. J. Wess and B. Zumino, Nucl. Phys. B70, 39 (1974)
3. For a review see
P. Fayet and S. Ferrara, Phys. Rep. 32 69 (1977)
P. Van Nieuwenhuzien Phys. Rep. C68, 189 (1981)
4. A.B. Balantekin, I. Bars, F. Iachello, Phys. Rev. Lett. 47, 19 (1981); Nucl. Phys. A370, 284 (1981); Yale preprint YTP-82-11 to be published in Phys. Rev. Lett.
5. A. Salam and J. Strathdee, Nucl. Phys. B79, 477 (1974)
6. T. Banks, A. Schwimmer, S. Yankielowicz, Phys. Lett. B96, 67 (1980); I. Bars and S. Yankielowicz B101, 159 (1981); I. Bars, Phys. Lett. B106, 105 (1981); Phys. Let. B114, 118 (1982); Nucl. Phys. B280 (1982); Yale preprint YTP 82-84, in Proc. Of Rencontre de Moriond 1982;
A. Schwimmer, Rutgers preprint RU-81-49.
7. I. Bars, to be published.
8. Y. Ne'eman, Phys. Lett. 81B, 190 (1979) and Tel Aviv preprint TAUP 134-81.
9. C. Becchi, A. Rouet, A. Stora, Commun. Math. Phys. 42, 127 (1975); J. Thierry-Mieg and Y. Ne'eman, Ann. Phys. N.Y. 123, 247 (1979); R. Delbourgo and P. Jarvis, J. Phys. A. Math, Gen. 15, 611 (1982).

10. P. Fayet, Phys. Lett. 69B, 489 (1977); 70B, 461 (1977);
S. Dimopoulos S. Raby, Nucl. Phys. B192, 353 (1981);
M. Dine, W. Fishler, M. Srednicki Nucl. Phys. B189, 575 (1981);
S. Dimopoulos, H. Georgi, Nucl. Phys. B193, 150 (1981);
S. Weinberg, Phys. Rev. D26, 287 (1982).
11. I. Bars, G. Veneziano and S. Yankielowicz, to be published.
12. S. Deser and B. Zumino, Phys. Lett. 62B, 335 (1976); D. Z. Freedman, P. Van Nieuwenhuzien, S. Ferrara, Phys. Rev. D13, 3214 (1976).
13 G. 'tHooft, in "Recent Developments in Gauge Theories", eds. G. 'Hooft et al. (Plenum, New York, 1980).
14. A.B. Balantekin, J. Math. Phys. 22, 1149 (1981) 22, 1810 (1981); 23, 1239 (1982).
15. A.B. Balanteksin, J. Math. Phys. 23, 486 (1982).
16. I. Bars, B. Morel, H. Ruegg, (CERN preprint TH 3333 (1982) to be published in J. Math Phys.
17. I. Bars and M. Gunarydin, CERN preprint TH 3350 (1982) to be published in Comm. Math. Phys.
18. V.G. Kac, in Differential Geometrical Methods in Math. Phys. ed. K. Bleuler, H.R. Petry, A. Reetz (Springer, Berlin 1978).
19. Y. Ne'eman, S. Sternberg, Proc. Natl. Acad. Sci. USA 77, 3127 (1980); P.H. Dondi and P. Jarvis, Z Phys. C4, 201 (1980); J. Math. Phys. A14 547 (1981); M. Scheunert, W. Nahm, V. Rittenberg,
J. Math. Phys. 18, 155 (1977); M. Marcu, J. Math. Phys. 21, (1980);
J. Thierry-Mieg and B. Morel, Harvard preprint HUTMP 80/8100.
20. H. Weyl, Classical Groups, (Princeton, U.P. Princeton, 1946)
21. A.M. Perelemov and V.S. Popov, Sov. J. Nucl. Phys. 3, 676 (1966) and 5, 489 (1967).
22. See e.g. W.G. Mckay and J. Patera, Tables of Dimensions, Indices and Branching Rules for Representations of Simple Lie Algebras (Marcel Dekker, N.Y. 1981); B.G. Wybourne, Symmetry Principles and Atomic Spectroscopy E.B. Dynkin, Am. Math Soc. Tr., Ser. 2, 6, 353 (1957); H. Freudental, Indag. Math. 16, 490 (1954).

INDEX